CNTs for the Next Generation of Electronic Devices

A Practical Introduction

Published 2026 by River Publishers
River Publishers
Broagervej 10, 9260 Gistrup, Denmark
www.riverpublishers.com

Distributed exclusively by Routledge
605 Third Avenue, New York, NY 10017, USA
4 Park Square, Milton Park, Abingdon, Oxon OX14 4RN

CNTs for the Next Generation of Electronic Devices: A Practical Introduction / R. G. Abaszade and A. K. Haghi.

Routledge is an imprint of the Taylor & Francis Group, an informa business

ISBN 978-87-438-0820-6 (paperback)
ISBN 978-87-438-0822-0 (online)
ISBN 978-8-743-80821-3 (ebook master)
ISBN 978-87-438-0824-4 (ePub)
DOI: 10.1201/9788743808213

CNTs for the Next Generation of Electronic Devices

A Practical Introduction

R. G. Abaszade

Azerbaijan State Oil and Industry University, Baku, Azerbaijan

Azerbaijan University of Architecture and Construction, Baku, Azerbaijan

A. K. Haghi

Department of Chemistry, Institute of Molecular Sciences, University of Coimbra, Coimbra, Portugal

River Publishers

Routledge
Taylor & Francis Group
NEW YORK AND LONDON

Contents

Preface

This book explains how to apply unique substances to solve the challenges of energy conversion and storage, along with improving efficiency. It introduces nanoengineered materials along with innovative methods to boost the performance of both green energy and technology. In this compact volume, readers will gain knowledge on applications of new nanoengineered materials such as those having a strong surface-to-volume ratio with more flexible electronic arrangements and stronger structure, and better heat and electricity conduction. It gives a comprehensive and coherent summary of recent advances and trends in how green energy can be used to our environment.

This book provides comprehensive information on how carbon nanomaterials contribute to clean energy and how novel nanostructured materials are applied in fuel cells. The applications of new types of nanomaterials to control environmental pollution effectively are introduced.

R. G. Abaszade

A. K. Haghi

Acknowledgments

The authors would like to acknowledge the help of everyone involved in this project and, more specifically, to our colleagues and reviewers that took part in the review process. Without their support, this book would not have become a reality.

Our sincere gratitude goes to the students who contributed their time and expertise to this book. We wish to acknowledge the valuable contributions of the postgraduate students regarding the improvement of quality, coherence and content presentation of chapters.

R. G. Abaszade

A. K. Haghi

Editorial Advisory Board

List of Abbreviations

LIBs	Lithium-ion batteries
GBNs	Graphene-based nanomaterials
TEM	Transmission electron microscopy
FTIR	Fourier transform infrared spectroscopy
SEM	Scanning electron microscopy
EDX	Energy-dispersive spectroscopy
MWCNTs	Multi-walled carbon nanotubes
SWCNTs	Single-walled carbon nanotubes
SWNTs	Single-walled nanotubes
TC	Thermal conductivity
CB	Carbon black
DSSCs	Dye-sensitized solar cells
LLZO	Lithium lanthanum zirconium oxide
PVDF	Polyvinylidene fluoride
VRFB	Vanadium redox flow batteries
CAES	Compressed air energy storage
GNPs	Graphene nanoplatelets
PSCs	Perovskite solar cells

About this Book

The main purpose of this book is to investigate and explain in detail the development and applications in nanoengineered green energy systems and technologies. Since energy demand across the world is on the rise, using sustainable and renewable energy is essential. Nanotechnology makes it possible to solve current issues in conventional energy systems (thanks to its excellent electrical, thermal and mechanical constituents). Postgraduate students interested in materials science, nanotechnology, chemistry, energy engineering and environmental science will find comprehensive expertise in this book. This book explains the applications of nano-designed materials along with their performance and sustainability in green energy systems. This book covers a wide range of subjects reflecting the most recent developments and new trends in the industry, providing:

- Synthesis, electronic layout, green design and work in energy storage and production are described for carbon nanomaterials in the scope of green energy systems.
- The importance of carbon nanostructures as they are essential parts of fuel cells.
- Advanced nanomaterials with particular applications on lithium-ion batteries.
- New studies on nanoengineered materials for solar cells.

In addition to discussing the main concepts and synthesis methods, this book underlines how carbon nanotubes are applied in modern energy technologies. It is important to solve the issues related to the environment, the ability to make those materials in large quantities, and how long the materials will last. With the help of nanoengineering, this book seeks to boost the growth of energy solutions that are energy efficient.

R. G. Abaszade

A. K. Haghi

About the Authors

R. G. Abaszade is an associate professor at Azerbaijan State Oil and Industry University (since 2022) and Azerbaijan University of Architecture and Construction (since 2024). He is the author of 153 articles, nine textbooks, 11 subject programs and six inventions, and papers in refereed journals with over 933 citations and h-index of 18, according to the Google Scholar database. Dr. Abaszade received a degree in physical electronics (Bachelor) at Baku State University in 2003, optoelectronics (Master) at Baku State University in 2006, and PhD in physics in 2018. In 2019, he created the "Innovation Research" department at the International Academy of Ecoenergetics and continues to research the purchase, analysis and applications of nanomaterials. Their research is carried out in the framework of joint cooperation with the scientific centers of the USA, Germany, Belarus, Kazakhstan, Turkey, Russia, Ukraine, Uzbekistan, UK, Hungary, Pakistan, Poland, Portugal, Mexico, Chile, China, India and Italy. They are engaged in complex work related to the acquisition, addition, analysis and applications of graphene-based samples. At the same time, they create various nanoelements (negatrons, diodes, capacitors and transistors) based on graphene-based materials. The research group is continuing experimental work in the direction of obtaining various samples, studying them and preparing nanomatrix devices based on new smart technology for the implementation of smart city and alternative energy supply. Dr. Abaszade is editor-in-chief of *Ecoenergetics* and is a leading reviewer on the editorial board of numerous journals. In 2019, he became a professor at the International Academy of Ecoenergetics and, in 2023, he was an invited professor at Nursultan Nazarbayev University. Dr. Rashad Abaszade is the director of the Turan International Research Institute.

A. K. Haghi is a retired professor who has written, co-written, edited or co-edited more than 1000 publications, including books, book chapters and papers in refereed journals. He holds a BSc in urban and environmental engineering from the University of North Carolina (USA) and two MSc degrees, one in mechanical engineering from North Carolina State University (USA) and another in applied mechanics, acoustics and materials from the Université de Technologie de Compiègne (France). Professor Haghi was awarded a PhD in engineering sciences at Université de Franche-Comté (France). His extensive educational background and supervisory roles underscore his expertise and contributions to the field of engineering sciences. He is an appointed Honorary Research Associate (HRA) at the University of Coimbra, Portugal, and a regular reviewer of leading international journals. He holds two UK design patents.

CHAPTER

1

Carbon Nanomaterials for Green Energy Systems

1.1 Introduction

The subject of nanoscience is progressing and evolving. Carbon layers were prepared in the shape of forming tubules first by Iijima in 1991 (Abaszade et al., 2021, 2023a). Carbon nanotubes (CNTs) are cylinders made by twisting graphene sheets. CNTs are available as single-, double-, multiple- and thin-walled. One-dimensional cylindrical structures, called single-walled carbon nanotubes (SWCNTs), have a proto-filament that is much thinner than it is long. Double-walled CNTs are made of two nested SWCNTs.

According to Abaszade et al. (2020b), you need two CNTs to come together or one to be absorbed by another if you want multi-walled carbon nanotubes (MWCNTs). Although the distance between its walls is very small at 0.34 nm, the tube's entire diameter is more than 10 nm.

Goyal et al. studied carbon nanomaterials to help improve how well lithium-ion batteries and supercapacitors hold up, in order to improve their functioning and raise their efficiency (Goyal et al., 2024). Later, Tsai (2024) improved the anode in lithium batteries, and presented different approaches. The special structure and proper porosity of carbon-based nanomaterials (CBNMs) seem promising in future technologies.

Apart from energy storage, CBNM is needed in energy conversion systems. The discovery of 2D carbon material monoethylene glycol (MEG) opened up a new horizon in energy technology that makes use of moisture in the air

DOI: 10.1201/9788743808213-1

(Wei et al., 2024d). With this awareness, we can use CBNMs and rely on energy from recyclable natural resources. Yang et al. (2023f) and Freitas et al. (2024) used biomass to help produce nanomaterials, which means we can safely consider waste materials as a source of energy.

The classic green economic entity, biomass CBNMs has the potential for deployment in attractive renewable energy technologies. Given their possible applications in sustainable energy technologies, green methods for CBNM production also trigger tremendous interest. Since they are based on biomass and waste materials (Sezali et al., 2024; Yadav & Daniel, 2023), green principles for CBNMs have lately attracted researchers since the synthesis of these materials may be done without negative consequences towards the environment. Moradi et al. (2024) reported the application of the functionalized carbon nanostructures (CNTs, graphene and their derivatives) to facilitating the development of flexible electronics.

This chapter presents crucial opinions on the material-structure-function relationship and offers guidelines on the future innovations in flexible technologies of carbon basis.

1.2 Synthesis and Characterization of Carbon Nanotubes

If you consider how CNTs are made, you can find three main branches of study. As CNTs are found in many different areas and have a lot of unique features, many people are focusing on their research and manufacturing. This section will discuss how today's scientists are creating, processing and characterizing CNTs. By rolling a single sheet of graphene into a tube, you can see how CNTs get formed along the atomic axes.

The nature of the shared vector plus the position and arrangement of the tube have an impact on the electronic structures and the spectral properties of geminal (neighboring) carbon. That is the reason why SWCNTs exhibit chirality, which is a twisting of the graphene sheet as it circles the central axis. Adjusting the chiral angle makes it possible to get metal-like or semiconductor-like conductivity one-dimensionally. Diameters of SWCNTs (Figure 1) are generally between 1 and 2 nanometers (nm). It may be stretched to any length, reaching hundreds of nm to tens of micrometers. Individual SWCNTs made properly have at worst one structural flaw every 106,000 carbon atoms along their few micrometer lengths.

Furthermore, theoretical configurations depicting a double-walled CNT as single-walled or a multi-walled exist. Indeed, with the right catalysts, your product gains double-walled nanotubes with weight percent exceeding 95%.

Direct printing of the tubes could be significantly more efficient than any method presently in use to produce microscopic transistor components (Figure 2).

Table 1 Synthesis methods of carbon nanotubes.

No.	Method	Temperature (°C)	Catalyst Used	Advantages	Disadvantages
1.	Arc Discharge	4000	Transition metals (Fe, Ni)	High crystallinity, MWCNT/SWCNT possible	High energy, low purity, complex setup
2.	Laser Ablation	~1200	Co, Ni	High-quality SWCNTs	Expensive, low yield
3.	Chemical Vapor Deposition (CVD)	500–1200	Fe, Co, Ni on substrates	Scalable, controllable growth, low cost	May require purification
4.	Plasma Enhanced CVD (PECVD)	300–700	Ni, Fe	Vertically aligned CNTs, lower temperature	Complex plasma system
5.	Flame Synthesis	~1000	Iron-based salts	Cost-effective, continuous growth	Poor control over chirality and purity
6.	Floating Catalyst CVD (FC-CVD)	800–1200	Ferrocene, thiophene	Continuous growth, high yield	Gas handling safety concerns
7.	Electrochemical Deposition	Room to moderate	Template-assisted	Simple setup, site-selective growth	Mostly limited to MWCNTs, template removal step

Figure 1 Cylindrical twisting of graphene layer and illustration of a single-walled carbon nanotube.

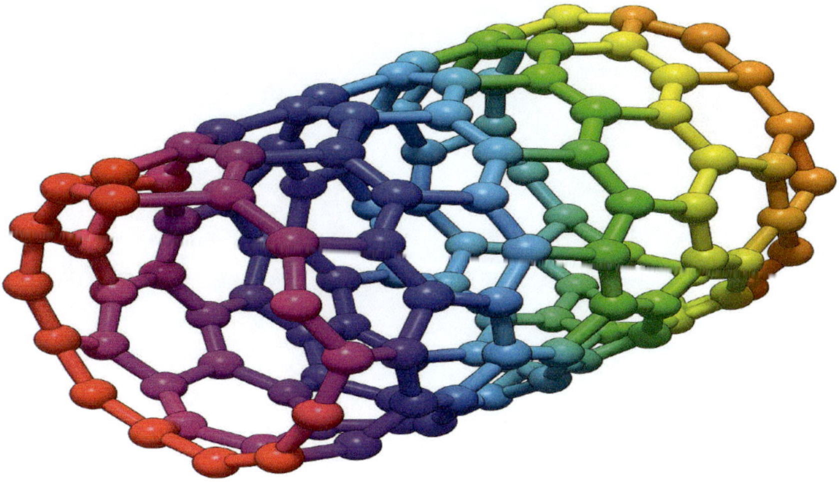

Figure 2 Schematic of a double-walled carbon nanotube.

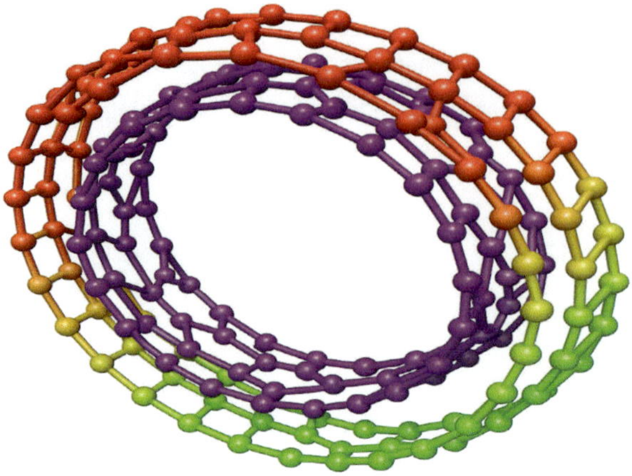

MWCNTs (Figure 3) are interesting since they are composed of various kinds of flaws and additional layers. The exterior diameter of a nanotube might measure only a nanometer, or it might go to 10 nm, and their length can exceed several micrometers. About 1–2 nm is the width of its inner space.

Figure 3 Multiwalled carbon nanotube.

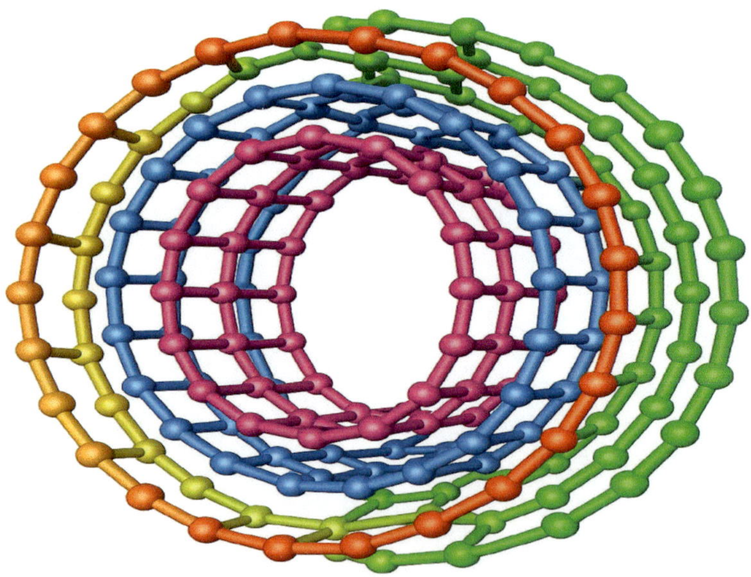

Sometimes these layers are arranged as coaxial, cylindrical, coil or cross-section coils. It is also possible to have fullerenes resembling nanotubes, but with both open and closed ends. There isn't always a clear winner between them since supplements come in most forms and the results are good in both cases. Initially, fullerenes and nanotubes were obtained by exposing graphite to either an electric arc or laser. Subsequently, the gas vapor was condensed to prevent reactions with other chemicals. Thus, there are several tested methods to make CNTs, but every method has its own drawbacks.

1.3 Arc Discharge Method

Usually, two pure graphite electrodes are required for arc discharge, as shown in Figure 4(a). The anode mainly consists of graphite, or it can be made with graphite and other metals. At this stage, the graphite powder is combined with the metals and placed in the center anode's hole. Thin contact between the electrodes quickly sets off in an arc. This chemical reaction happens at low pressure (ranging from 30 to 130 tor or 500 tor) and when mixed inert gas or reagent gases are present. Electrodes are caused to come closer together slowly, until the strength of the current reaches around 50-150 A. When the gap between the electrodes reaches a high temperature, the carbon vaporizes from the anode and gets burned. That being the case, the separation between the anode and the cathode must be maintained at a set distance. This process results in forming plasma between those structures. The gap between the electrodes is set using voltage (25-40 V) control, so the plasma keeps running for as long as needed. Some reactions end quickly while others need much more time.

The reactor produces a variety of things. There will be (1) a dark and sticky soot on the reactor's wall, (2) a spider web pattern between the cathode and the inner wall, (3) a grey mound protruding from the cathode, and (4) a spongy ring surrounding the grey deposit. It is common to use Fe, Ni, Co, Mo, Y or a mix of them to make the alloys. Catalysts made of two different metals gave slightly better results. What's produced is amorphous carbon, metal particles coated with carbon, and carbon in polyhedral shape.

If carbon and the metal catalyst are put into a DC arc discharge, multi-walled nanotubes (MWNTs), functionalized multi-walled carbon nanotubes (FMWNTs), GNP, functionalized graphene nanoplatelets (FGNP) and microplastic and nanoplastic particles (MNP) form inside the deposit. These are formless soft soot particles made up of MWNTs, FMWNTs and SWNTs. Most SWNTs have diameters between 1.1 and 1.4 nm and their lengths can be many microns. Almost all the collarette material is SWNT or SWNT in bundles. However, this happens only if there are special catalysts present. Arc discharge makes use of heat, catalyst type and gases; there are some physical and chemical variables. These value factors determine the growth and size of the nanotubes as well as the type.

Figure 4 (a) Arc discharge apparatus and (b) experimental device in liquid N_2.

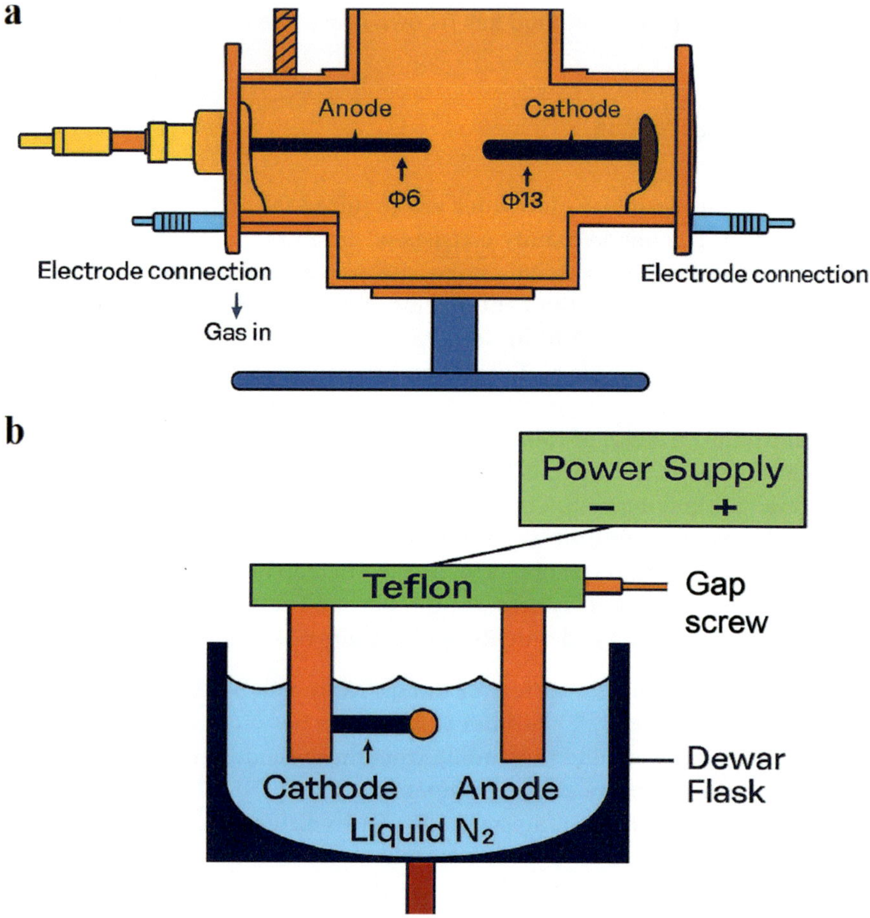

1.4 Laser Ablation Method

The laser ablation approach transforms graphite into vapors at high temperatures using a pulsed laser (Lahoz et al., 2020). MWCNTs and SWCNTs can be created by this standard practice. As part of SWCNT production, whose diameter is around 1-1.6 nm, tiny metal particles functioning as catalysts must be situated in graphite. It is a method that makes use of arc discharge. A lot can play into the output of CNTs, including many elements that may impact the production. The properties of catalysts, how much energy the laser emits, the surroundings' temperature and pressure, the type of noble gas, and the way carbon is introduced influence the result. It is clear that the laser sends its light mainly to the layer made of 1.2% Co/Ni and 98.8% graphite in the composite. The reaction is done inside a quartz tube oven at 1200°C upon introducing argon gas as the carrier at 500 tors. As a result of the introduction of the argon gas, the air around the walls draws the warm steam near the enclosure walls into the cold area where argon is present. The next step involves releasing gas and silted carbon onto the walls that are formed into gas and create CNTs. The result of using laser ablation resembled the purifier CNTs made during arc discharge. The output in Raney and Li–Ce catalysts changes under the influence of temperature, catalyst amounts, and laser light power (see Figure 5).

Figure 5 Schematic of the laser ablation method.

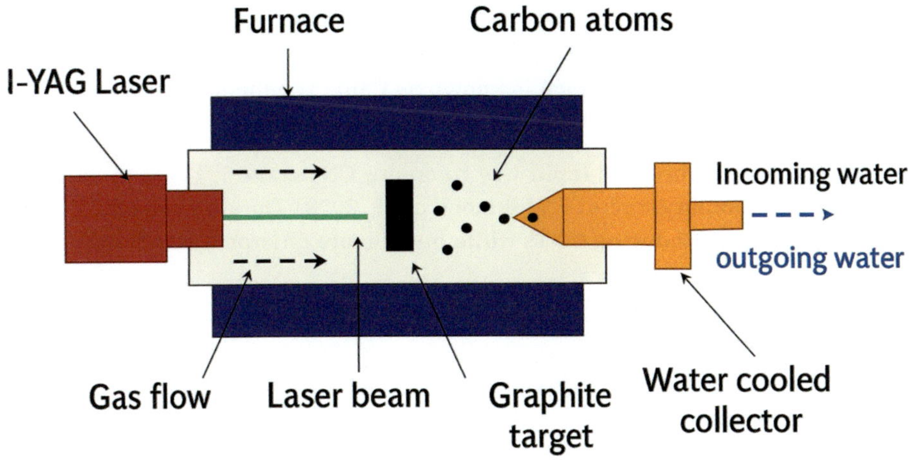

1.5 Chemical Vapor Deposition Method

Consider, for instance, CVD in which moving hydrocarbons gas through a quartz tube with hot sources or microwave plasma creates a carbon network on the reactor's inside material. Then, the reaction gives energy to a carbon atom in the form of a gas.

First, the hydrocarbons are heated, and this process produces carbon by forming hydrogen carbon bonds. Next, carbon changes into metal (nickel, cobalt, iron, or a combination) catalyst particles that attach to the metal (nickel, cobalt, iron, or mixed) catalyst substrate. The right conditions lead to the formation of CNTs. CVD is helpful for producing wafers since it provides the little power needed, same temperatures, mild purity, and allows wafers to be enlarged. Moreover, this well-known method is responsible for both MWCNTs and SWCNTs (to form SWCNTs, the temperature has to be higher than for MWCNTs).

Carbonaceous gas (including methane, carbon oxide, or acetylene) passes through a quartz reactor tube that uses an energy source (such as a microwave plasma or heated filaments) in the process. The heat released during burning goes to gas carbon atoms. First, using heat, hydrocarbons change into hydrogen carbon bonds and carbon molecules. In this kind of reaction, the carbon will be over the hot metal (nickel, cobalt, iron or their mix) catalyst particles that are placed over the substrate. This is the point at which all the elements come together. Nucleation happens in CNTs to give the best result.

CVD benefits are that it uses very little energy, temperatures are moderate, and the impurities are of medium quality, yet the main advantage is that profile dimensions can be easily increased. It can also lead to SWCNTs by using a higher temperature than the one needed for MWCNTs. Recent studies have focused on different methods, one of which is known as flame fragment deposition (FFD) and relies on liquefied petroleum gas (LPG).

Figure 6 describes this rapid way to create CNTs, and it helps to obtain a well-defined tubular structure (Mahmood et al., 2023). Duraia et al. (2024) used other options for producing CNTs while being more environmentally safe.

1.5.1 Functionalization techniques

CNT helps composite systems by taking part in functions besides just controlling the production of formula. BI@MWCNTs showed that the properties of MWCNTs–CO_2H changed after the addition of a benzimidazole moiety. Various methods—surface area test, DLS, SEM, and FT-IR—proved that the BI@MWCNTs were prepared and characterized successfully. To find the main outcomes, the adsorption performance of BI@MWCNTs against Pb^{2+} and Cd^{2+} in pH ranges from 2.5 to 10 was looked into.

It was found that the easiest way to get rid of the two ions was at a pH level of 5. In light of the findings, BI@MWCNTs were better at attracting Pb^{2+} ions than Cd^{2+}. Adjustments were made with the help of contact time, metal concentration, and an amount of BI@MWCNTs to improve the adsorption.

Figure 6 Schematic for catalytic chemical vapor deposition making CNTs using methane.

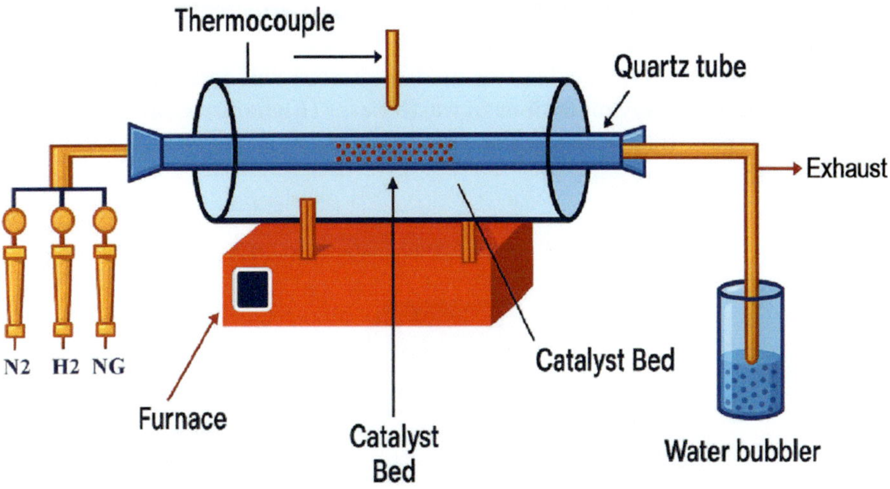

If sensors are used and the process lasts 20 minutes, both Pb_{2+} and Cd_{2+} will be reduced to 0 at a concentration of 50 mg/L (Pb_{2+} and Cd_{2+}), with 6 mg of BI@ MWCNTs and at a pH of 5.0 and a temperature of 25°C.

A lengthy adsorption rate test was conducted to check how the time for adsorption influences and reports its rate. Research also revealed that the ions were adsorbed differently and quicker depending on the place of adsorption on the BI@MWCNTs material. The other form of the kinetic equation was found from the amount of adsorbed adsorbate on BI@MWCNTs surfaces.

Langmuir isotherms model's were observed to show even closer results to those found in the Pb^{2+} and Cd^{2+} adsorption from BI@MWCNTs with various initial concentrations and at the selected temperatures (25 °C and 35 °C). This system's monolayer was fitted with the models developed by Langmuir and Freundlich. It was found that at 25°C, ΔG, ΔH, and ΔS equaled the following amounts. Since the value of ΔG was negative, this proves that both Pb^{2+} and Cd^{2+} absorbed naturally. Even though applying energy and heating make adsorption easier, the fact that the interactions between the solid and liquid increase as they become more disordered suggests that this process is still taking place.

During the first five runs, the sorbent was able to remove the same percentage of contaminants; afterward, the percentage decreased just slightly. Results from the studies reveal that the adsorptionable Pb^{2+} and Cd^{2+} using the produced BI@MWCNTs was set at 100% and 98%. On this basis, BI@MWCNTs are able to absorb more metals cations when compared to other carbon-based materials. In addition, BI@MWCNTs can be recycled and applied for six circular cycles, which means they are both economical and effective in the treatment of wastewater and removing the metals (Pb^{2+} and Cd^{2+}). It has been found that adding heterocyclic chemicals after acid treatment will raise the nanotubes' ability to take up heavy metal ions in waste water (Elghamry et al., 2023). Also, the polymer part tied to the CNTs may make them stronger and keep them steady during stress (Figure 7).

Figure 7 Functionalization methods in CNTs.

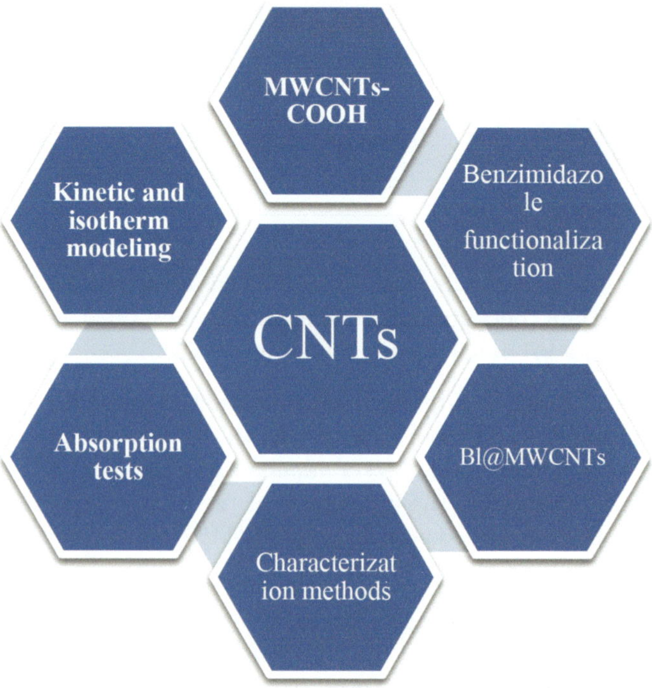

1.6 Characterization Techniques

Understanding how CNTs function in structure, mechanical, and electrical terms requires knowledge of them to be gained through characterization. Spectral Energy Distribution (SED), TEM, and X-ray diffraction (XRD) are some of the common methods to analyze objects. The CNTs, referred to as MWNTs, were produced in the size range of 10–30 nm and their lengths varied. Later, a coat of uniform magnetite nanoparticles (with a diameter of 15–35 nm and a spherical shape) was applied to them. Diffraction considerations show that nanotubes are created by the combination of carbon and nano-sized magnetite. After scanning the electron microscope, it became easy to decorate the CNTs. On the basis of thermal degradation, the transit of a substance can assist in controlling the process since both types of nanotubes show two mass losses where the first mass loss in the decorated nanotubes has an activation energy of 6.75 kJ/mol and the second mass loss has an activation energy of 25.74 kJ/mol. Decomposition of decorated CNTs causes two different mass losses. Mass transport regulation is confirmed, as the value of apparent activation energy for the initial mass loss is found to be 20.61 kJ/mol. In relation to the activation energy of 79.54 kJ/mol, the second mass loss highlights that chemical reaction played the main role in this phase of thermal decomposition and that temperature gave the most control to it. Because their outer layer is highly decorated, decorated CNTs are better able to resist being damaged by high temperatures, so they are more stable thermally. With these techniques, we are able to check the shape, geometry, and crystal structure of the made CNTs. As an illustration, SEM found CNTs bound to particles and, in turn, this made it possible to use energy storage devices (Olcay et al., 2024). In addition, using X-ray diffraction, Raman scattering, and the method of studying the coherence length, CNTs produced by arc discharge and gadolinium-doped CNTs were explored. In line with the Raman spectra, there is a surge of the Gd-CNTs' spectrum to higher frequencies at their maximum absolute value. It is clear from X-ray and Raman spectra that inclusion of gadolinium greatly increased the intensity of the peaks. Why doped nanotubes have better intense peaks is mainly because their CNTs contain fewer defects in their structure. As an example, the insertion of gadolinium into CNTs brings out a noticeable contrast change (Abaszade et al., 2020b; Figure 8). Besides, tests using electrochemistry are used to check the quality of CNTs in various devices, including sensors and batteries. Here, this study avoided the costly and complicated method of synthesis and succeeded through the simple two-step method. The process of producing MnO_2/CNT nanocomposites was tested using XRD and Field-emission Scanning Electron Microscopy (FESEM) tools. The samples of CNTs/MnO_2 and MWCNTs/MnO_2 showed the characteristic structure of porosity. Electron transport and performance at the electrochemical level were highest with the composite of MWCNTs and MnO_2. Additionally, microwave-assisted synthesis takes less time to prepare than chemical ways. From electrochemical examinations, it seems that Visualization for Electronic and Structural Analysis (VESTA) of MWCNTs/MnO_2 will find many applications in electrochemical sensors and energy storage devices. As CNTs play a key role in electrochemical movements (Yousf et al., 2024), tests such as cyclic voltammetry and impedance can be carried out to check if the structure and surface modification are suitable.

Figure 8 Raman spectra of pure and gadolinium-doped carbon nanotubes.

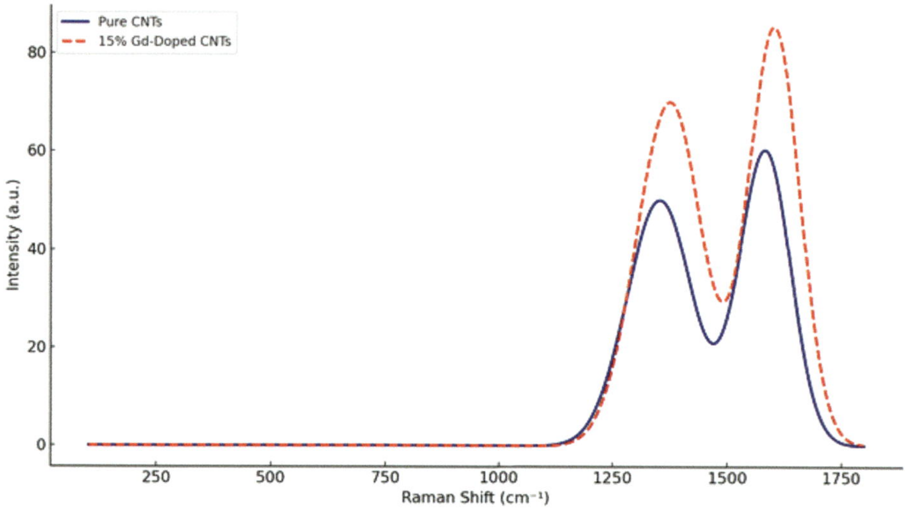

It was found that the light sensitivity of CNTs in λ = 400–900 nm varies with the applied voltage U (Van = 1, 3, 5, 7 and 9 volt). Maximum F_0 and the greatest sensitivity range appear when U = 1 V. Ip has its greatest value of about 0.2 μA when the wavelength is λ~715 nm. Whereas other networks have four experimental points per wavelength, specific networks with six wavelengths were created (λ = 640 nm, λ 720 nm, λ 740 nm, λ 760 nm, λ 780 nm, and λ 880 nm), each with its own dataset and matching curve, for the samples of nanowires tested. It has been realized by researchers that fourth-order polynomials should be used for characterizing the curves. There is only a little photosensitivity in the I-V curves studied using the operating voltage range 1-9 V at λ~720 nm (Figure 8).

Figure 9 illustrates that sensitivity of the carboxyl group-containing CNTs is different at different wave lengths (between 400 and 900 nm) and voltage levels (1 V and 3 V). All the current tests measured peaks between λp = 1600 mm and λp = 5875 nm, with the highest values belonging to λp = 2700 nm, λp = 3760 nm, λp = 4840 nm and λp = 5875 nm (see Figure 10). Figure 9 illustrates the sensitivity of CNTs to light of different wavelengths and voltages. It can be seen in Figure 10 that the photocurrent at U = 1 V is uniformly distributed over the wavelengths from 600 nm to 875 nm and is stronger than those at the peaks λ of 600 nm, 700 nm, 760 nm, 840 nm, and 860 nm. When the wavelength is λp4~840 nm, the greatest amount of current change (photosensitivity) is 0.1 μA.

Figure 9 CNT light sensitivity spectrum versus bias voltage.

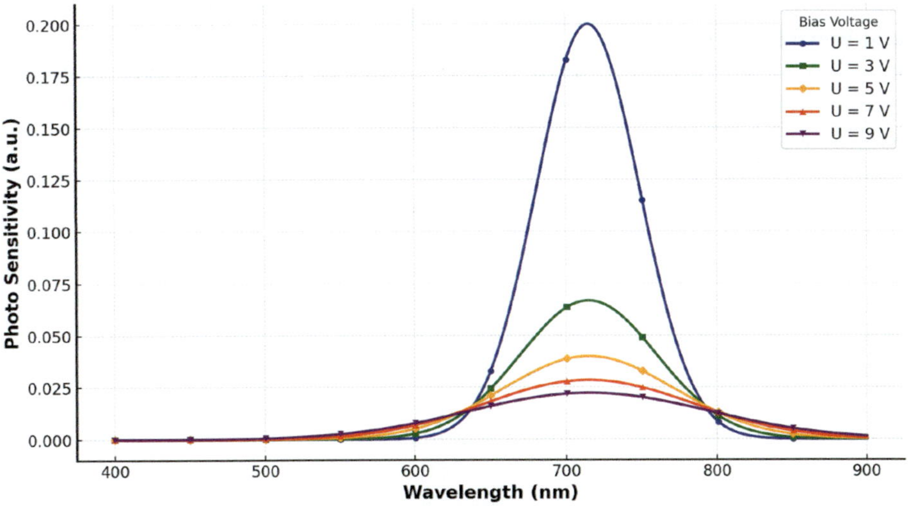

No photosensitivity is seen at 400-575 nm, and all other response peaks have photocurrent under 0.1 μA. U = 1 V shows that the range between 400 and 900 nm is classified as 'low intensity, light-sensitive.' At 3 V, I explanations found the peaks in photocurrent at wavelengths of λp1~400 nm, λp2~460 nm, λp3~540 nm, λp4~620 nm, λp5~700 nm, λp6 ~820 nm, and λp7~875 nm. Because the studied nanomaterial has a high photocurrent of Ip6~1.93μA, the peak of photocurrent at λp6~820 nm moves to the low wavelength region. No sensitivity to light was noticed in the range λ = 475–525 nm. In this case, the non-photosensitive range has been set at zero. The UV-NIR photosensitivity of the functionalized carbon nanotube (fCNT) due to different biases (U = 5 V, U = 7 V and U = 9 V) is displayed in Figure 10. At λp = 1400 nm, λp = 2460 nm, λp = 3540 nm, λp = 4600 nm, λp = 5675 nm, λp = 6720 nm, λp = 7760 nm, there were five volts of electric current. Photocurrent achieves its highest value at λp = 6720 nm, with peak current of Ip = 62.67 μA, but peak currents of the third (λp3540 nm) and fourth (λp = 4600 nm) peaks are λp = 4600 nm and Ip = 41.05 μA, respectively. When performing this sedimentation, the zero-photo sensitivity band lies from λ = 800 to λ = 900 nm. The highest photocurrent intensity was noticed for λp = 1400 nm, λp = 2460 nm, λp = 3600 nm, λp = 4635 nm, λp = 5740 nm, λp = 6800 nm, λp = 7840 at U = 7 V. Maximum photoconversion happened with an input of Ip = 62.04 μA and a λp = 6800 nm wavelength. Another important step happened when the photocurrent was less than 0.5 μA. It was shown at waveband λ = 475–575 nm that there was no sensitivity to light. When the experiment was done using 9 V and a wavelength range of λ = 520-620 nm ($\Delta\lambda$ = 100 nm), the spectra came out weak and had a modest photocurrent peak at λp = 2560 nm. Palladium oxide (PdO) has photocurrent peaks Ip = 31.14 μA at wavelength λp = 3640 nm and Ip = 51.03 μA at λp = 5800 nm. Photocurrents of 0.7μA at λp = 1475 nm and 0.58μA at λp = 4680 nm result when the maximums are reached. Higher voltages up to 9 V (U) may be applied to increase the sensitivity of the investigated fCNT sample, as seen from photocurrent spectra analysis. The sample couldn't be used for light detection since its response to light (photocurrent) was low. The standard voltages for fCNTs is 3 V when the laser light ranges from 400 to 900 nm, and 5 V when the light covers 400-800 nm. For the best sensitivity, the device should be set for the highest voltage, that is U = 5 V. Under 2 V bias and λ = 720 nm incident wavelength, Figure 10 indicates that the photocurrent is Ip = 2.67 μA, suggesting that fCNTs may detect light at different wavelengths applied. Comparison of the photosensitivity curves between fCNTs and pure CNTs proves that the former is more sensitive. Clearly, the highest parasitic absorption in CNTs is Ip = 2.67 A for those with carboxyl functionalization, while pure CNTs have Ip = 0.185 A as reflected by the red line. With 14 times sharper sight of light than before, the fCNTs improve performance. I-V curves assist in changing how dependent the experimental spectra are on the voltage levels, showing the effects of light in fCNTs (Figure 10).

Figure 10 Photosensitivity spectra of U = 1, 3, 5, 7, and 9 V.

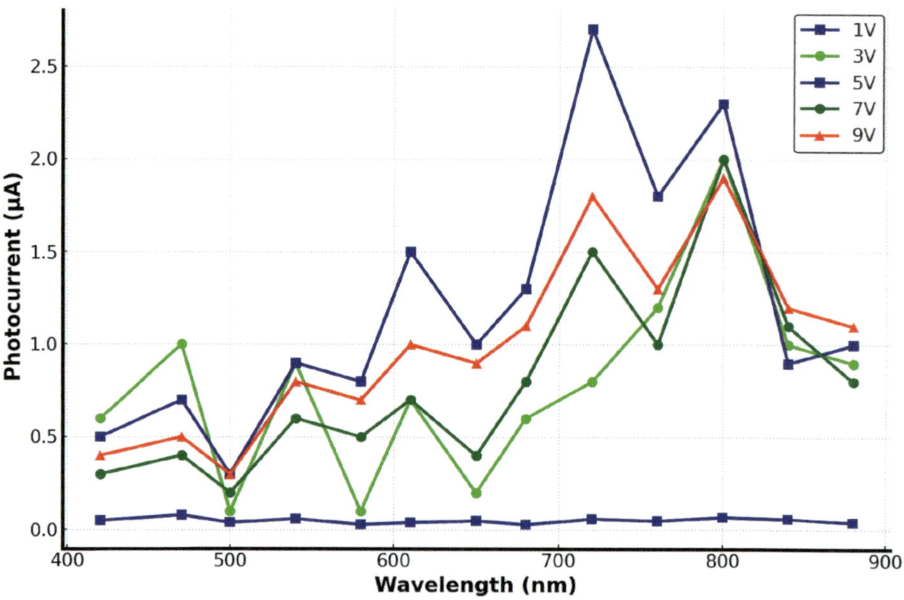

Using the arc discharge technique, we were able to make graphite, CNTs, and functionalized CNTs with carboxyl. SEM and EDX tools indicate that the materials measured worldwide are within a 100 nm range. How graphite's molecule is shaped depends largely on the carboxyl groups. From X-ray phase investigation, it is apparent that crystals in the produced graphite have a hexagonal structure. The samples are quite amorphous, though the presence of functionalized carboxyl groups makes them useful. These structures require us to look at them through deeper research (Stetsenko and Abaszade, 2023; Figure 11).

Figure 11 XRD analysis of graphite, CNT and FCNTs.

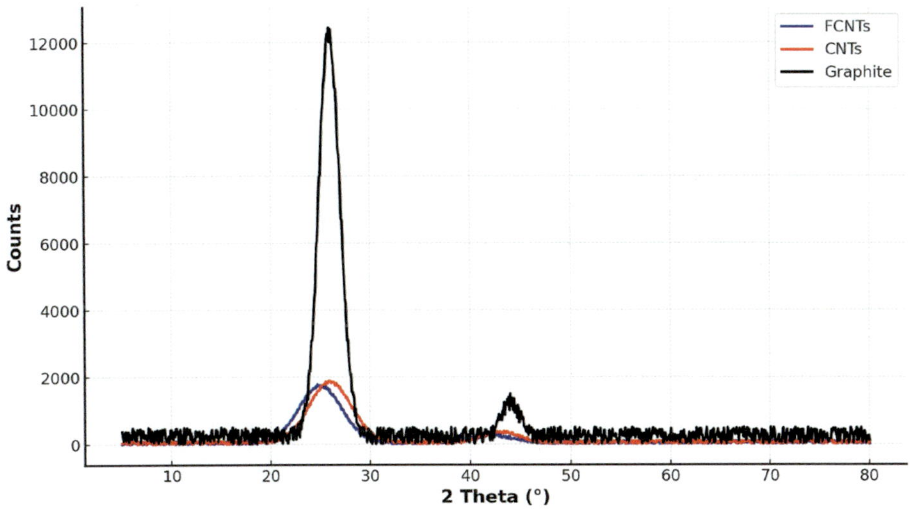

Figure 12 SEM analysis of (a) MWCNTs and (b) gadolinium-doped MWCNTs.

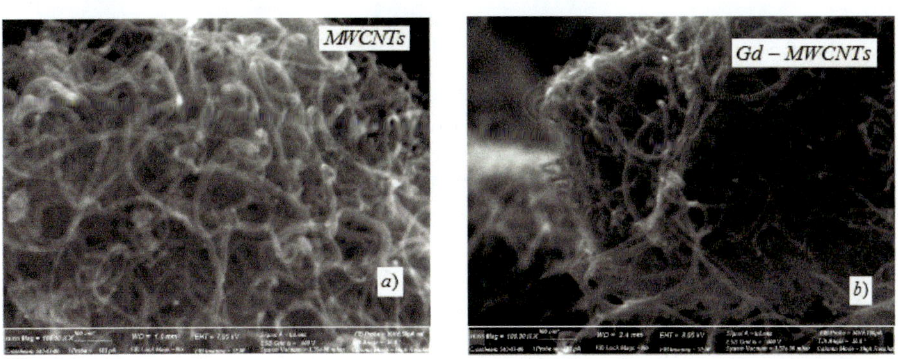

EDX analysis of pure MWCNTs shows that there are very few silicon elements along with carbon (Figure 12(a)). Moreover, element analysis reveals that C, O and Gd elements appear on the surface of the Gd-doped MWCNTs due to oxygen coming from the Gd ions (Figure 12(b); Table 1). Silicon forms in the ceramic containers through the process of treatment. As there is 9.5 mass % of oxygen to gadolinium, clusters of gadolinium oxide largely cover the MWCNTs (Table 2).

Table 2 EDX data of elemental composition of MWCNTs and Gd/MWCNTs samples.

No.	Sample	C (Weight %)	O (Weight %)	Gd (Weight %)
1.	MWCNTs	100	0	0
2.	Gd/MWCNTs	90	9.5	0.5

From Figure 13, the presence of surface function groups was investigated with FTIR spectrum for both MWCNTs and Gd-MWCNTs. Bands at 3400 cm^{-1} and 1610 cm^{-1} are connected to the stretching and bending of groups containing OHs in the alcohol. Lack of these peaks shows that there are no surface hydroxyl groups in Gd-MWCNTs. The peaks at 1350 cm^{-1} in MWCNTs are explained by the OH groups that are present.

Figure 13 FTIR spectra of Gd-MWCNTs and MWCNTs.

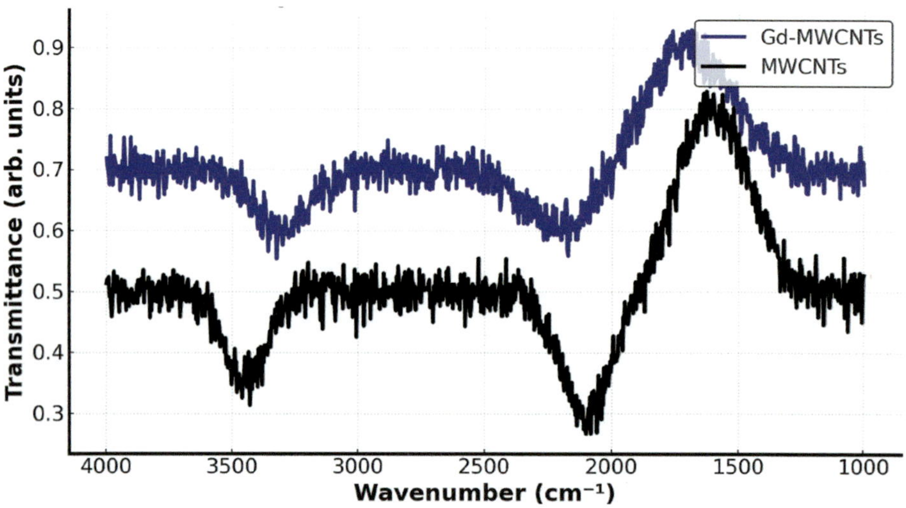

An important step before CNTs can be used in nanotechnology is making them and learning about them. Due to their scalability, CVD is the main way of producing CNTs. Nevertheless, the quality of CNTs is higher when they are produced by means of arc discharge or laser ablation. Many approaches have been used to find how CNTs are structured, composed chemically, and what their properties are electrically, so their usefulness can be explored. To carry out CNT research, there are many methods to produce them, attach new traits to them, and examine them. Many more applications of CNTs are being reported as a result of ongoing studies.

1.7 Electronic Structure of CNT

Taking the two kinds of CNTs (SWCNTs and MWCNTs), plenty of their electric properties are made better by their chiral and diamagnetic properties. Mostly, the band type in a CNT (whether it is metallic or semiconducting) is determined by its chirality. Depending on the chiral vectors they have (n,m), some SWCNTs become metals and others become semiconductors (Figure 14). The mentioned adjectives indicate the expected electronic band gap, as shown by Al-Dayel et al. (2024) and Shadidi et al. (2024).

Figure 14 Comparison of the bandgaps of metallic and semiconducting single-walled CNTs.

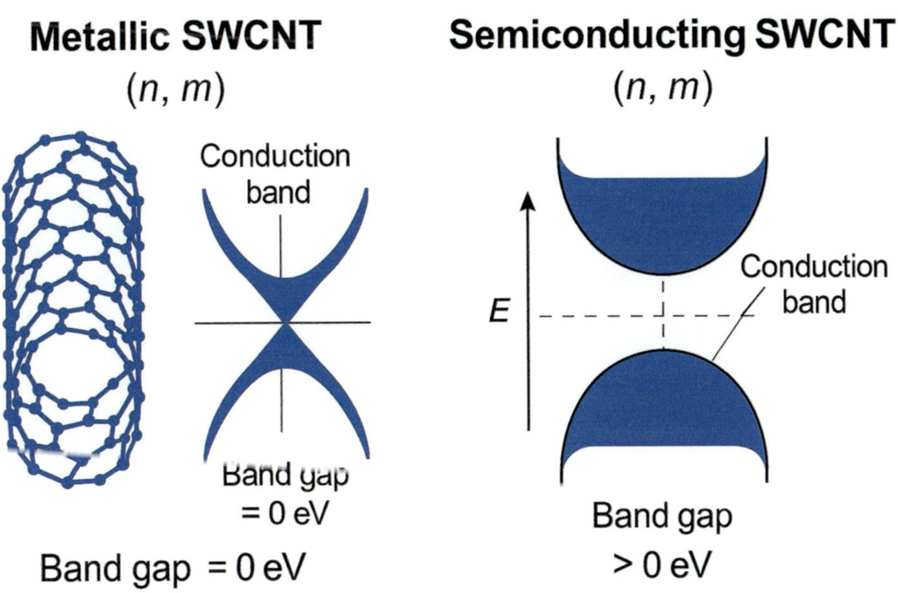

CNTs have unusual electrical properties since they are made by rolling graphene. The manner in which CNTs are formed and their size makes them act like metals or semiconductors. CNT's presence in electronics and its role in them vary according to its chirality. Generally, armchair, or metals, does not work as a semiconductor, however, zigzag and chiral tubes have proven to do so (Al-Dayel et al., 2024; Shadidi et al., 2024; Figure 15).

Clearly, the chirality expressed by the chiral indices (n,m) influences the electrical properties of SWCNTs. The main structural types of CNTs are explained by the way groups of carbon atoms are bonded.

- Since they are metallic in nature, armchair CNTs (n = m) are applied in situations that require high electrical conductivity.
- According to whether n,m can be divided by 3, the zigzag CNTs labeled as (n,0) may behave as a metal or as a semiconductor.
- Because chiral CNTs in the (n,m) form with n≠m and n≠0 are the most semiconducting and handleable, they play a key role in making sensors and transistors.

Figure 15 Zone structure of CNTs with armchair, zigzag and chiral shapes.

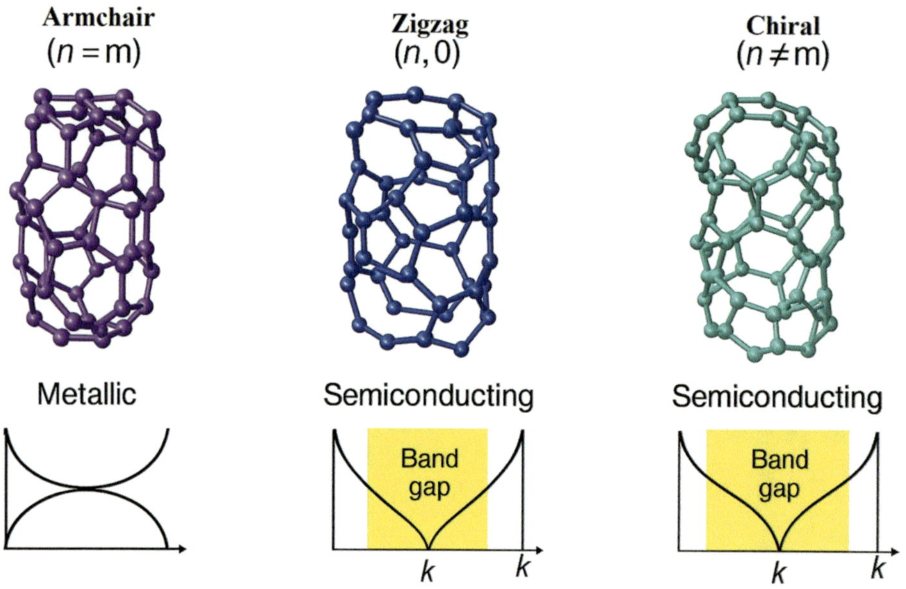

The way the band structures appear shows the electrical nature of the various regions in semiconductors and are also described as regions of green color. Overlapping conduction bands make up the valence bands, which have zero band gap, for metallic CNTs. Those semiconductive CNTs where the valence band is above the conduction band must spend energy to allow electrons to move. Since the way CNTs are used in nanoelectronics, energy storage, sensing and various other applications depends on this property-structure relationship, it is important to control the chirality when making CNTs. The electrical structure of a CNT made from rolling graphene is just like that of graphene:

$$E(k) = \pm\gamma_0\sqrt{1+4\cos\left(\sqrt{\frac{3k_z\alpha}{2}}\right)cos\left(\frac{k_y\alpha}{2}\right)+cos^2\left(\frac{k_y\alpha}{2}\right)}$$

where γ_0 is nearest-neighbor hopping energy (~2,7 eV), α is lattice constant of graphene (~2,46 eV), $k = (k_z,k_y)$ is wavevector.

Chiral vector (defines CNT geometry):

$$C_h = na_1 + ma_2$$

where (n,m) is chiral indices (defines CNT type: armchair, zigzag, chiral) and a_1, a_2 is graphene lattice vector.

Tube diameter is:

$$d_t = \frac{C_h}{\pi} = \frac{\alpha}{\pi}\sqrt{n^2+m^2+nm}$$

Allowed wavevectors around circumference (quantization condition) is:

$$k_{per.} = \frac{2\pi q}{C_h}, \quad q = 0, \pm1, \pm2, \ldots$$

Energy dispersion relation for CNT subbands is:

$$E_q\left(k_{par.}\right) = \pm\sqrt{1+4cos\left(\sqrt{\frac{3\alpha k_{par.}}{2}}\right)cos\left(\frac{\pi q}{N}\right)+4cos^2\left(\frac{\pi q}{N}\right)}$$

where $k_{par.}$ is wavevector along the tube axis, q is subband index, N is number of unit cell around the circumference.

A CNT is metallic if:

$$n - m = 3p, \quad p \in Z$$

Energy bandgap of semiconducting CNTs is:

$$E_g = \frac{2\gamma_0\alpha_{CC}}{d_t}$$

where α_{CC} is carbon-carbon bond length (~1,42eV), is tube diameter.

Density of States (DOS) for 1D CNT is:

$$DOS(E) \propto \sum_n \frac{1}{\sqrt{E-E_n}}$$

This leads to Van Hove singularities, which are sharp peaks in DOS for CNTs.

Alternatively, Van Hove singularities in electrical characteristics appear in the band structure of CNTs. The basic stories cause these flaws because the characters are made equally important. Enormous differences in electronic transport and DOS happen to CNTs because of these singularities. It is very important to understand the way electrons are placed in the energy levels using DOS and that control has an effect on CNTs and their electronic behavior. CNT functionalization switches the metallic form of CNT to a semiconductor one (and vice versa). Functionalized tubes and the chirality of the tubes play a big role in dividing the DOS. Armchair CNTs automatically have high density of states at the Fermi energy and thus they are metals; chiral CNTs, in contrast, might be metallic or form a gap with semiconducting properties. This section mainly investigates attaching various groups to CNTs (4, +4), (4, +2) and (4, +1) forms. The presence of right- and left-handed chiral structures greatly matters in DOS calculation. When nitrogen is introduced to CNT, the number of electrons that can participate in chemical reactions goes down, whereas the OH-group has a greater impact on CNT's DOS reduction than NH_2-group owing to the higher electronegativity of oxygen than nitrogen. We should remember that when the diameter of the nanotube is below 4 Å, the curvature of the short diameter cuts DOS changes caused by functionalization to the same value as the long diameter changes. According to many studies, when an external functional group is added to a CNT, it affects the highest occupied molecular orbital (HOMO) and lowest unoccupied molecular orbital (LUMO) of the pure CNT and, consequently, the energy gap. Even though researching the influence of Electron-Donating Groups (EDGs) on CNT DOS would be helpful, this article concentrates mostly on Electron-Withdrawing Groups (EWGs). In addition, it also explored how the functionalization process affects double-walled carbon nanotubes (DWCNT). It is hoped that CNTs can improve or offer new opportunities for electrical, energy, and catalytic applications thanks to their special properties. Substituting carbon atoms with other atoms, placing functional groups on the outside of CNTs, and stuffing the CNT channels have for a decade been found to be the effective methods to change the electrical property of CNTs. This has made CNTs develop different electrical properties, which has increased the possible uses of CNTs. Among all the methods, filling the CNTs with various functional materials stands out as the main way to update the properties of the nanotubes because it leaves the carbon core untouched.

In this book, we look into the ways filled CNT heterostructures are created and how they are used. The internal structures in the heterostructures are well described by using the most advanced HR-TEM and AC-TEM approaches. Due to being constrained in space, CNT can arrange certain chemicals with an organization unique from the bulk and are modulated by diameter. Sensitivity to this is the CNT's electronic property. Lots of metastable nanomaterials have been prepared by nano chemical reactions in nanotubes, suggesting that the CNTs confined-hollow-space can be used to make new materials. Besides, these enhanced techniques, Optical Absorption Spectroscopy (OAS), Raman, and X-ray Absorption Spectroscopy (XAS) offer easy and efficient ways to study the impact of foreign compounds on nanotubes' electronic features.

P and N doping appeared when the samples were treated with enough donor and acceptor to reach saturation. Thus, the devices using CNTs are made using the existing knowledge and methods to affect their electrical behavior. Even though CNTs have been combined with different materials over the years, certain issues stop these structures from becoming widely used. Since the filling process is influenced by different outside factors, the outcome is inconsistent and might reduce the quantity of the product. Therefore, getting good quality samples involves using more helpful and improved filling techniques. The yield of CNT filling may be influenced by CNTs' size, the selected filling substance's characteristics, the way of filling, and the surrounding environment. In some instances, the amount made is not enough for the needs, but sometimes it goes above and beyond 100%. Paying attention to both the required uses and characteristics of the filled CNTs is essential regarding pragmatic issues in the filling. If CNTs are filled with catalysts they will be used as supports, then both the filler and the filling process need to ensure strong catalytic activity, along with a homogenous spread of the catalysts throughout the CNTs. There are some practical outcomes that change the way the organization works. As an example, when using solvent to fill CNTs, the filling process could block some of the solvent inside, thus affecting the qualities of the material. The features of heterostructures have not been studied in great detail. Work on the sample's structure, electrical features, and spectrum is done subsequent to the sample being prepared in a single tube. It is also hard to make field-effect transistors with structure made from purified semiconductor substances and CNTs. When a substance is put into the CNT, both the weight and surface charge distribution change, so the initial separation conditions are no longer important. Nevertheless, if CNTs are loaded with other materials, those materials could leak out and cause weak spots when they are serviced in vacuum conditions or liquids. Getting the CNT walls to attract certain chemical groups might stop the visitors from leaving the interstitial region and help to keep them in the CNT. A further method is to have the bigger composite placed over the CNT, then use the encapsulation approach—logo or shield—to make CNTs more suitable for electrical devices (Tyagi & Negi 2023).

Raman spectra of CNTs are commonly used to explore the electronic structure because they express much about the vibrations and electronic state of these nanotubes. Both the G and D bands in Raman spectroscopies can clearly distinguish sp^2 bonded carbon atoms and those with defects. According to these bands' intensities, you can deduce whether the CNTs are high quality and good conductors. A petroleum gas from Iraq was used to create CNTs successfully. Flames were used to help deposition for synthesizing CNTs. A number of methods were used to study the materials and also to check their similarity to a standard. The supply of MWCNTs from Sigma Aldrich was 95% pure, but Raman analysis demonstrates that a fraction of FWCNTs were found therein. SEM and TEM photographs show signs that nanotubes are being formed. That is normally the case since examples with more defects usually have lower conductivity (Mahmood et al., 2023). It can be summed up as their electrical characteristics depending on how the tubes are chirally arranged along with their size and what materials contact them. One has to discover advanced applications such as nanoelectronics, sensors, and energy-storage material before utilizing the special features of CNTs. CNTs have incredible electrical features because of their structure and the molecules around them. Metals or semiconductors can be applied for nanoelectronics, sensors, and storing energy.

1.8 Novel Applications of Carbon-based Nanomaterials

Graphene and its related forms are well studied for being highly valuable and finding multiple applications in different sectors. Due to graphene's ability to fight against both Gram-positive and Gram-negative types of bacteria, more research efforts are being placed on this material and similar compound like graphene oxide (GO) and reduced graphene oxide (rGO). As it has excellent surface-to-volume ratio, the development of graphene gives us a new biocidal nanomaterial that doesn't easily degrade and also provides good mechanical and surface properties. The bond between graphene and cells is explained by graphene's operating segments, edges, and faces. Bacterial reactions to graphene-based nanosheets depend on their sheet size, how much surface area they have, their roughness, hydrophilicity, dispersibility, and functionalization. It was found in studies that exposing graphene-coated masks to UV light sterilizes them and causes a significant increase in temperature. Studies involving graphene membranes that do not let a virus through are just starting. Regardless of graphene-sheets' usefulness in improving masks because of their known ability to kill bacteria and viruses, new methods should still be sought for safe handling of filters containing nanomaterials (Lavorato and Fontananova, 2023). Experts have applied graphene and its varieties in environmental protection for the purpose of purifying water and capturing pollutants (Figure 16).

Figure 16 Novel applications of carbon-based nanomaterials.

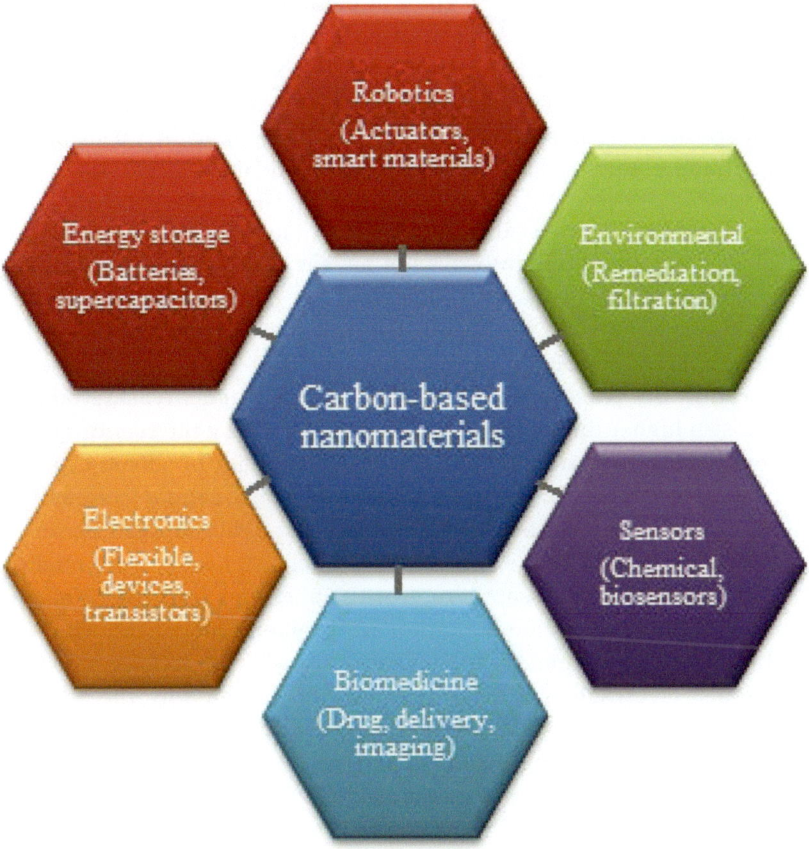

Here we will discuss how GO is used to remove metal ions and organic matter from water and the latest techniques used in water treatment. Many aspects impact GO's adsorption performance, namely pH, temperature, and how many bonded oxygen groups are present. Currently, most studies are carried out in a lab setting, so the use of GO in cleaning water is not expected to happen right away. More investigation should prove that the theory performs well in engineering applications. In the future, scientists will try to boost how GO and its mixes work in complex water solutions and check how stable and recyclable they are, aiming to increase their usefulness in actual water treatment engineering. Involvement of these materials in water treatment is becoming more important due to progress in research and technology. Moreover, if the construction and types of chemicals added to complex water are matched with various pollutants in various water sources, the adsorption abilities will become more reliable and adaptable to the kinds of pollutants present. Coming up with new methods that lower the expense of preparing materials is an important area for study. In the end, GO and its composites have a huge impact on water resources and environmental protection when used in water treatment (Liao, 2024). Its special qualities allow graphene to be applied in various sorts of digital equipment, like sensors, transistors, and bendable electronics.

In wet transfer for chip-scale study, there are many problems dealing with the water because the graphene does not have a supporting carrier substrate, which is harder at a larger scale. When there is more bonding strength, the quality and graphene yield get better with the semidry transfer method. But, as-grown graphene is usually better than semidry-transferred materials if all goes well in the process, and the uniformity of graphene achieved by wet transfer is still higher due to the extra stresses caused by the semidry transfer. Metal contamination is greater than the tolerance limit of silicon CMOSs for these processes. Moreover, hotspots appear because copper contamination locally affects only parts of the graphene sample, and this happens because the stirring process is not possible. Under ideal circumstances, fiber-on-lightpath (FOL) design could include sapphire growth substrates since these may help to decrease high copper contamination in graphene monoliths. The use of thermal release tape in semidry processing causes CO emissions and changes the materials' electricity-handling abilities. The reason why the CO– bond part is more polluted is that the polar component in the X-ray Photoelectron Spectroscopy (XPS) is more obvious in this region. By measuring the higher frequency waves, it was observed that the resistivity of wet-transferred graphene was much lower (from 450 to 550 Ω sq^{-1}) than that of semidry-transferred graphene (ranging from 1000 to 1650 Ω sq^{-1}). The findings from the four-point measurements were the same as the results mentioned earlier (wet-transfer is 100–400 Ω sq^{-1}; semidry-transfer is few kΩ). In short, wet transfer leads to more graphene, fewer carbon impurities, and better electrical results than does semidry transfer. One issue with wet transfer is that graphene is often damaged as a result of pressing or touching it even slightly, since it can be hard to manage after copper is etched.

On the other hand, half treated wafer carriers allow handling the wafers without robotic setup. Even though both techniques have good and bad points that can be addressed with more careful work and better equipment, it is believed that semidry will give a clear advantage in practical situations and for high production (Wittmann et al., 2023). To sum up, many technologies have heavily relied on graphene and other related nanomaterials. Because of their unique features, they make it possible to perform new activities in electronics, medicine, energy storage, and making the environment cleaner. Still, for these materials to be most effective, new research and development are vital to deal with manufacturing, up-scaling, and including them in present-day technologies.

1.9 Sustainable Carbon-based Nanomaterials

CBNMs have attracted attention from energy researchers since they promise to play a big role in developing advanced energy technology. Because of their characteristics linked to nanoscience, such as vast surface area and being eco-friendly, bio-nanomaterials are attracting much interest nowadays. This section gives a summary of recent progress in nanocellulose and CBNMs for batteries and supercapacitors. This section discusses the chances and the obstacles of producing nanomaterials with biomass. Other conductive nanoparticles brought up in this section are MXene which may pose a potential risk to the market of bio-nanomaterials for energy storage (Sezali et al., 2024). With whispering-gallery-mode (WGM), we are able to give populaces free clean water and also make unique generators that create sufficient energy for their needs. Even so, it creates a big obstacle for storing hydrogen, despite its very appealing advantages. A better way to handle hydrogen is to find effective ways to keep it as clean fuel for tomorrow. Storing hydrogen using nanomaterials or in nano-form is simple, light, safe, and can give efficient hydrogen fuel whenever needed. This section describes several solid-state nano materials such as activated carbon, carbon fibers, metalorganic frameworks, covalently linked frameworks, nano porous organic polymers, and metal hydrides. Besides, advancements in hydrogen fuel cell technology have been made for both stationary and portable equipment. A recent study (Manzoor et al., 2024) addresses key achievements thus far, lists main areas for improvement, and gives tips on what is forthcoming in sustainable energy (Manzoor et al., 2024).

Carbon nanoparticles for use in commercially accessible and commercialized products in the biomedical sector constitute another environmentally favorable material. Because of their biocompatible character and special optical qualities, these materials have been investigated for drug delivery, biosensing, and cancer treatment. Naturally acquired carbon quantum dots, explored for years as bio-imaging and therapy tools, are proof of the healthcare industry's green effort. Many agriculture, energy, environmental protection, and medical issues can be resolved with the help of carbon-based nanocomposites. Since these units have specific qualities and can be produced in environmentally friendly ways, they will play a role in building a more eco-friendly world.

1.10 Energy storage/Generation Applications

Thanks to nanotechnology, electric equipment can be made smaller. Their performance is dependent on the way the structure of the supercapacitor (SC) electrodes, the size and type of pores, the surface area, and the presence of functional groups is set up. Commerce prefers porous carbon as an anode because it is low in cost, nonharmful to humans, and ready for use. When nitrogen atoms were introduced to porous carbon, this made the material's properties better, resulting in a rise in supercapacitor performance. Because it has a big external surface, high particular capacitance, excellent retention, high aspect ratio, and remains in two dimensions, this material performs better. Because of nitrogen's doping action, carbon materials look promising for use in energy-related applications. What's more, using synthesis, materials can be designed with controlled pore sizes, shapes, and how easily other molecules can enter them. Because they exist in thin 2D sheets, Nickel-Cobalt-Manganese (NCM) can be used as electrodes for supercapacitors. Yet, the energy output of nitrogen-doped carbon materials isn't sufficient enough and it is necessary to enhance composite or hybrid products with additional nano-micro substances for commercial use. According to the latest studies on nitrogen-doped porous carbon and various heteroatom-doped carbon-based all dielectric self-supporting (ADSs), it is possible to enhance their super capacitive properties through correct material design (Zafar et al., 2023).

Over the past few years, considerable advances in the use of graphene-based nanomaterials (GBNs), especially in solar cells, energy storage, and hydrogen production technology, were made. The exceptional properties exhibited by graphene in electrical, mechanical, and thermal matters qualifies it to be used to improve the effectiveness and stability of solar cells. A schedule has been produced by Bagade et al. (2023), and it shows that graphene-enhanced solar cells do not perform worse; on the contrary, they perform better and are much more resistant. Likewise, Bist et al. (2023) emphasized why the stability of perovskite solar cells (PSCs) is enhanced by graphene and graphene derivatives. Graphene as anode materials in the field of energy storage have been shown to enhance the capacity of lithium-ion batteries and stability in terms of cycling. Bi et al. (2023) made a detailed study of graphene anodes and its beneficial effect on the battery. Bhardwaj et al. (2024) demonstrated how porous networks of graphene could be fabricated on polyaniline electronically and how they operated as excellent electrochemical supercapacitors. In addition, Byun et al. (2024) examined the process of degradation in lithium-ion batteries and focused on how graphene can suppress its impact. Graphene-based materials have also been useful in hydrogen production and storage. A solar-thermal tool of hydrogen generation was proposed by Bardwell et al.

(2024), and the hydrogen storing ability of nitrogen-doped reduced GO with decorations of palladium nanoparticles was demonstrated by Boateng et al. (2023). The study offered by Bongu et al. (2024) reviewed graphene-based 2D material use in rechargeable batteries and in hydrogen-based technologies, challenging them and outlining future perspectives. Nanomaterials based on graphene show potential in the business of biomedicine. Their use in highly sensitive biosensors was studied by Baruah et al. (2024). Moreover, Benettayeb et al. (2023) demonstrated the possibilities of graphene-based adsorbents in the removal of harmful pollution in the environment and touched on dye-sensitized solar cells (DSSCs) with organic dyes and titanium dioxide monolayers (Bristy et al., 2024). All in all, these studies have demonstrated the wide range of usage of graphene in energy and biomedical applications, besides bridging areas of technological and economic challenges that still need to be overcome in order to utilize graphene-based solutions in a real-world setting.

Looking for new and eco-friendlier ways to save energy results from both the worsening of our Earth and the using up of fossil fuels. The reasons for this are the green qualities, plenty available in the environment, ease of getting, low production costs, and their ability to endure for a long time. Full instructions on how to design and build an energy module with supercapacitors were explained by the research. One study makes use of porous carbon electrodes (with a surface area of 44-1050 F/g) and biomasses that have surface areas of 215-3532 m^2/g. These electrodes are said to preserve 99.7% of their capacitive function after going through 1000 cycles. Between 5.1 and 138.4 Wh/kg is the energy level they can have. This work discusses the key parts of biomass and their effect on preparing porous carbon. It also looks into the effects of various porous carbon structures and tries to explain their reasons. Advances have been made in fuel cells, carbon capture, and finding uses for biomass nanoparticles. In this section, we study the recent developments in energy usage and storage by these materials. This review is aimed at supporting further advancement in energy storage research by giving details and predictions about the current state and future problems in creating biomasses based porous carbon composites. Carbon materials produced from biomass have been seen to be suitable for supercapacitors because of their safety and effectiveness when deployed (Saeed et al., 2024). Storing and generating energy is greatly helped by graphene-based materials that include graphene, GO, activated carbon, and CNT. These are some main ways in which they are used

1.10.1 Supercapacitors

Supercapacitors are one of the most attractive energy-storage tools, noted for their strong power, fast charge and discharge, and endurance. Based on studies, graphene and CNTs are chosen as the most appropriate carbon nanofillers for supercapacitors, while the use of activated carbon was already considered. Since they have a high surface area, excellent conductivity, and great strength—especially necessary for fast storage and transfer—these materials are highly respected. Graphene especially has caught the attention of researchers in the field of supercapacitors. What makes it unique as an electrode material are its considerable surface area and the conductivity it shows. By way of example, graphene cycles keep their excellent stability; this material created by self-propagating high-temperature synthesis has a loss of only 4% of its capacitance after a thousand cycles, as shown by Vozniakovskii et al. (2023). According to research on experimental data, supercapacitors made from self-propagating high-temperature synthesis (SHS) graphene show great potential. Regardless of having less specific surface area, samples of few-layer graphene (FLG) showed improved capacitance.

Even after we finished tests on these layered samples after 1000 cycles, we did not notice any noticeable loss in their strength. Thus, graphene with several layers offers a well-organized structure, equal particle sizing, and large capacitance values. In $LiClO_4$ electrolyte, a FLG anode managed to display a capacity of 265 F g^{-1} (0.2 mA g^{-1}). Capacitance values for the sulfonated carbon electrodes were twice or more than those obtained with similar carbon electrodes. Moreover, the electrode kept as much of its capacitance as after the first cycle of 1000 uses. Few graphite layers have been suggested for use as supercapacitor electrodes. Because the electrodes are now better designed and made, carbon nanomaterials are being used. The specific capacitance and energy density of such materials could be enhanced to a large extent since these hybrid materials form a composite structure. CNTs can survive for many years without falling apart.

It is important to mention that the low-impact character of the supercapacitor was due to the use of renewable feedstocks used in its production (Figure 17). It is also suggested a simple way of making activated carbon using spherical, flower-like sheets, and porous flakes from recycled face shield material. It is important to note that the specific surface area and pore volume depend greatly on the amount of activation temperature applied. When activated at 900 °C, the substance had a high specific surface area (1571 m^2 g^{-1}) and its mesopores had a diameter of 3 to 5 nm. Moreover, activated carbon ink and a specially designed coating process can be used to manufacture flexible supercapacitors with adjustable nano porous electrodes. We looked into how the performance of the flexible supercapacitor (Table 3) was affected by the Nafion binder's amount. The top electrochemical results came from activated carbon electrodes that were slot-die coated with 15% of Nafion. A high energy density of 21.8 $Whkg^{-1}$, a power density of up to 20,600 Wkg^{-1}, and rising cyclability 96.2% after 10,000 cycles when discharging at 2 Ag^{-1} were the main attributes of the activated carbon capacitor containing Nafion.

Figure 17 Comparison of capacitance in carbon-based nanomaterials for supercapacitors.

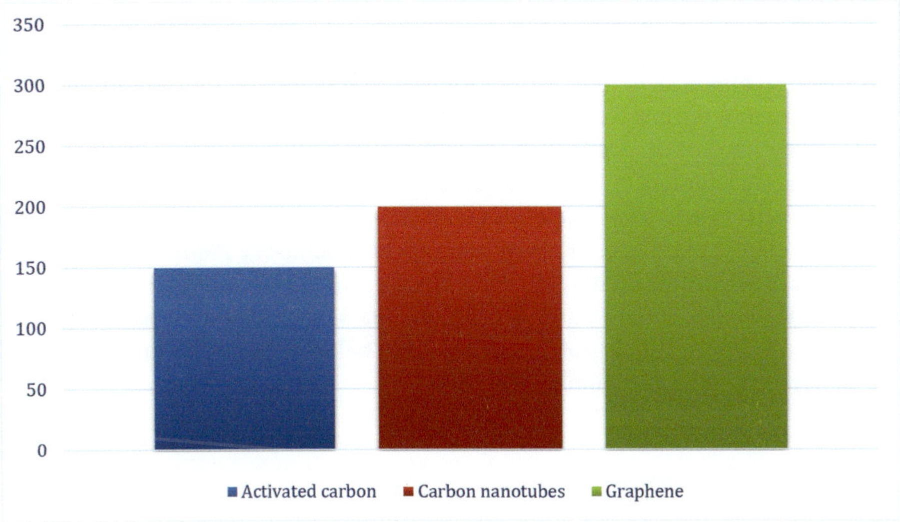

Table 3 Carbon-based nanomaterials in supercapacitors.

No.	Nanomaterial	Capacitance (F/g)	Feature	Application
1	Activated carbon	150	High porosity, moderate conductivity, cost-effective	Conventional supercapacitors
2	Carbon nanotubes	200	High conductivity, strong mechanical properties	Advanced electrode materials
3	Graphene	300	Very high conductivity, large surface area, superior performance	Next-generation high-performance supercapacitors

For this reason, this disposal method proposed for COVID-19 PET face shield plastics can be applied widely. Based on high promise for business, what is shown here is that utilizing large-scale wide-range slot-die coating is capable of making supercapacitor electrodes of AC. Reddygunta et al. (2024) used environmentally friendly materials to produce flexible supercapacitors, which marked the start of green energy storage technology. Even though carbon-based nanoparticles are good for supercapacitors, their production and use are still being improved. Experts are working hard to come up with new modalities to make supercapacitors strong and effective. Since engineered nanomaterials (such as graphene) have more energy, they are much more suitable for fast charge/discharge and quick power packages, mirroring the focus of new technology. The study by Preez and Halma (2024) achieves a critical review of applications, environmental behavior, and health risks of GBNs. Their research paper identifies common applications of GBNs in the field of electronics, energy equipment, biomedicine, and environmental restoration, due to unusual physicochemical properties of these materials. The authors are, however, concerned over their environmental non-degradation, human bioaccumulation, and toxicological impacts on both aquatic and terrestrial lifeforms. These important toxicity mechanisms are found in the oxidative stress, membrane disruption and inflammatory response in relation to exposure routes and material sizes, shape, and surface functionalization. It is noted that in practice standard risk evaluation procedures and long-term design solutions are necessary to achieve the balance between innovation and both environmental and human risk.

1.10.2 Lithium-ion batteries

Lithium-ion batteries (LIBs) are employed mainly in electric cars and small gadgets. Since being added to LIBs, graphene has recently led to a stronger performance with more energy density, more reliable cycle life, and speedier charging and discharging. Due to having multiple layers, being electrically good, and a high O{N}insertion surface, graphene is considered for both the LIB anode and cathode. One benefit of graphene in LIBs is that it helps enhance the performance of electrode materials (Figure 18).

To illustrate, the association of rGO with iron phosphates causes the iron phosphate particles to stay individual and boosts the lithium-ion transfer in cycling performance (W. He et al., 2024). It is evident from Figure 18 that introducing graphene-derived nanomaterials into lithium-ion batteries can offer numerous benefits, mostly related to increased electric conductivity, resilient support, and faster movement of ions inside the batteries. Therefore, lithium-ion batteries with added graphene in the electrodes have the potential to be used in future electric cars and electronics due to their strong life and speedy recharging. Because LIB are highly powerful and keep working for a long time, they make a good and valuable energy storage device. Since Li+ ions need time to move and the electrodes are not good conductors, today's batteries do not have a very high energy density. Nonetheless, because of this trade-off, nuclear power doesn't have the strength to compete with some clean types of energy. In addition, the way in which each section functions is not equal, which makes the battery's performance and life span unstable.

Figure 18 Graphene-based nanomaterials in lithium-ion batteries.

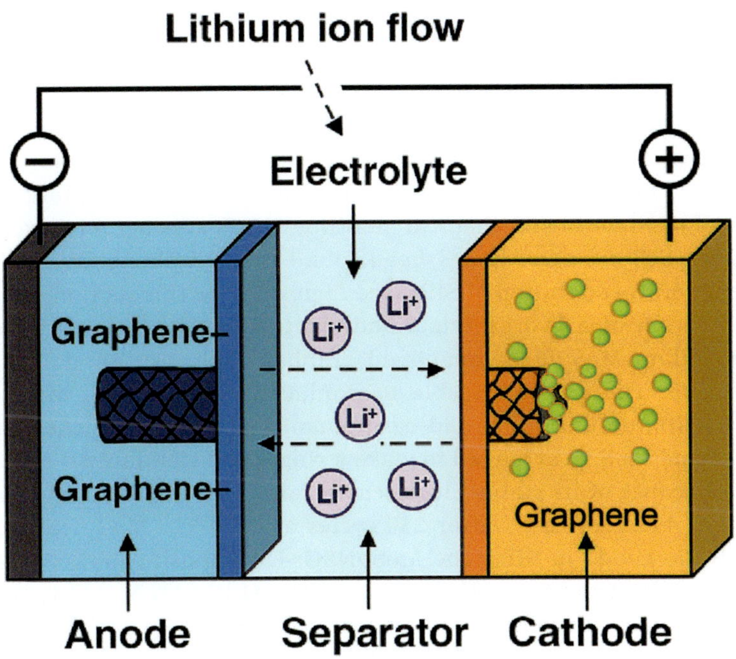

Therefore, batteries based on lithium-ions get the most out of anodes with added strength and better power qualities. Due to having a unique setup and strong electrochemical abilities, materials created from graphene are being tested as cathode materials for lithium-ion batteries. In general, there are much better standards in this field now. To this point, Alathlawi and Hassan describe the advances in graphene-based materials for lithium-ion battery anodes and cathodes. Their paper ends by mentioning the remaining problems and making suggestions for more studies. The current collector of cathode electrodes can be upgraded with graphene, since it brings both strong mechanical support and the improved electrochemical features necessary for making better lithium-ion batteries (Alathlawi & Hassan, 2024).

Although it is difficult to add graphene nanomaterials in commercial batteries, it can help improve their performance. Because of its interesting properties, graphene can advance the way energy is stored and delivered, which will play a leading role in future energy storage applications. With a graphene and carbon composite, lithium-ion batteries have the potential to store more energy. We are not yet at our energy storage goal.

1.10.3 Sodium-ion batteries

Because sodium is easy to access and not expensive, sodium-ion batteries (SIBs) may turn out to be the best option for LIBs. SIB performance is enhanced by creating new anode materials, among them being graphene and its analogs. This stems from their good electric conductivity, large surface space, and outstanding mechanical qualities, GBNs appear to be exceptional options for SIB systems. It has been found that graphene composites are beneficial for the operation of SIBs (see Figure 19). In this section, we explain the application of a hydrothermal and electrochemical process for making a free-standing MoS_2/homatom-doped graphene network anode for silicon ion batteries. The process is able to produce graphene with nitrogen and sulfur distributed uniformly and easily manageable. Each of the few-layers graphene and MoS_2 is arranged to make a single self-standing 3D structure in parallel. Because of its stable electrochemical properties and high reversible capacity of 443.9 mA h g^{-1} after 250 cycles at 0.5 A g^{-1}, MoS_2/NSGs-G suits very well the anode in SIBs. The MoS_2/NSGs-G composite showed great rate performance by providing 491.5, 490.5, 453.9, 418.1, 332.8, 333.1, and 294.4 mAh g^{-1} at current densities of 0.1, 0.2, 0.5, 1, 2, 5, and 10 Ag^{-1}. Furthermore, MoS_2/NSGs-G aided the speed of sodium ions at the interface and gave a major pseudocapacitive effect. Thus, MoS_2/NSGs-G could be a good anode to help boost the performance of SIBs.

Ding and Zhang (2023) use a first-principles density functional theory (DFT) surface model as a tool to explain the system mechanism of sodium intercalations at the nitrogen-dopants anode materials of SIB. In their study, various nitrogen doping structures such as pyridinic nitrogen, pyrrolic nitrogen, and graphitic nitrogen were systematically investigated as how each type of nitrogen doping contributes to an electronic structure, adsorption energy, charge transfer, and diffusion processes of sodium ions. The findings indicate that nitrogen doping plays a very important role in improving the absorption capacity and promoting the diffusion of sodium ions by altering the local electronic surroundings and inducing energetically advantageous intercalation locations. Of the doping types, pyridinic nitrogen has the best potential of boosting sodium storage capacity. Ding and Zhang (2023) is an excellent window into theory-based insight of high-performance graphene-based anode material development of next-generation sodium-ion energy storage systems.

Suggested by Ding and Pei (2024), a better active disturbance rejection controller (ADRC) design made specifically suitable for pumped storage hydro units has been introduced. There exist specific requirements of pumped storage hydro units as ones playing a significant role in large-scale energy storage and grid stabilization. The suggested approach involves improvements in traditional ADRC techniques in that it maximizes the tuning of control parameters as well as changing the architecture of the extended state observer (ESO). By means of vivid simulations and comparisons, the research reveals that the enhanced controller provides a faster dynamic response, lower overshoot, and higher control of uncertainties and external disturbances. Such improvements are associated with better and more economical working of the pumped storage systems that will enable them to be integrated with modern renewable-based power grids.

In accordance with Ding et al. (2024), there is a detailed examination of the latest developments in fast-charging anode materials in LIBs and the essential issues and levels of technology that are needed over the increasing demand of rapid energy storage. The paper critiques an extensive variety of anode materials alongside carbon-based ones, alloy-type ones, and conversion-type ones by taking into consideration their electrochemical properties, lithium-ion permeation, structural longevity, and surface/interface engineering techniques. Particular focus is paid to the processes leading to high-rate capabilities, e.g. pseudo capacitance and nanostructures, and the limiting impact of volume expansion and the instability of the Solid Electrolyte Interphase (SEI) layer. Other new trends that are mentioned in the review include two-ion transport, composite anode design, and machine learning-aided material design. All in all, the paper can be regarded as an effective guide to the future development of tho next-generation of fast-charging LIB technologies.

Energy storage and conversion technologies continue to develop in their partnership with novel materials and well-engineered system configurations. In the latest literature, one can find the concept of utilizing pumps as turbines in less than 0.1 megawatts of pumped hydro energy storage systems with enhanced efficiency and cost reductions (Laconi et al., 2024; Licheri et al., 2023). Pulsed laser ablation methods of the nanosecond scale have allowed the synthesis of nanoparticle zero-valent iron with designed surface characteristics, which are beneficial to the environmental and catalytic fields (Lahoz et al., 2020). LIB performance has been boosted by novel engineering of a nanostructured silicon anode that provides higher capacity and cycling stability (Lan, 2023). The materials based on graphene are still displayed, and the reduced GO/MXene hybrid electrodes have shown enhanced efficiency in vanadium redox flow batteries (VRFB). Exfoliated graphene composite membranes have been proven to show better ion selectivity and endurance (Pahlevaninezhad et al., 2023). Even green synthesis process has seen some potential in photovoltaics with examples including the green production of silver nanoparticle using plant extracts (Palanichamy et al., 2023). Kinetics of thermal decomposition of CNTs functionalized with magnetite nanoparticle have been demystified, which can help us understand their thermal management aspects and magnetic properties (Olcay et al., 2024). Reviews point out the development of thermal management techniques of lithium-ion batteries and emphasize a strong requirement of heat dissipation improvements to increase battery lifetime and safety (Ortiz et al., 2024). Research is also concerned with the problem of corrosion control with the involvement of CBNMs serving as a sustainable way to maintain the lifespan of materials in adverse environments (Pan et al., 2024). Together, these interdisciplinary innovations point to the vibrant development in the field of material science and engineering that is vital to address future needs in energy storage, environmental cleansing, and sustainable solutions.

Ding et al. (2023a) developed an insight into open-circuit voltage (Voc) losses in perovskite quantum dot solar cells (PQDSCs), which still make a significant obstacle to the full potential of photovoltaic efficiency. This is well analyzed by Ding et al. in terms of non-radiative recombination pathways, interfacial charge transfer losses as well as energy level mismatches to reduce Voc. The authors apply a mixture of experimental and theoretical characterization to illustrate both the effects of surface traps and ligand dynamics on their device, and how the extraction of charges may be inefficient. The means to alleviate Voc loss, e.g. surface passivation, optimized ligand design, and charge transport layer design, are presented in detail. The paper provides important understanding of PQDSCs core loss mechanisms as well as design guidelines to improve efficiency of PQDSCs in future technologies.

Advanced energy materials have experienced rapid growth as new nanomaterials and device geometries come to market to help improve energy collection, storage, and conversion efficiencies. Most recent research focuses on the custom design of porous carbon materials obtained through lignin, specifically propelled in the edge of high-energy lithium-ion capacitors, with

sp^2 and sp^3 hybridization bonding playing the essential role in performance insights (Peng et al., 2023). The new photovoltaic fields implement the innovative material platforms (chalcogenide perovskites and perovskite/silicon tandem cells) that improve the traditional restrictions on power conversion efficiency and current density (Pyun et al., 2023; Rai et al., 2023). Lithium-ion and all-solid-state battery technologies have been promoted by advancing silicon nanostructures, such as collecting high-strength anodes with better tap density to optimize capacity and stabilize cycling involving silicon (Qiao et al., 2024; Raman et al., 2023). The advancement in energy device performance using graphene and composite thereof is still on an upward trend on editing supercapacitors and fuel cells because of advanced composing techniques and design (Quintero et al., 2023; Rahman et al., 2023). Additionally, new electrospinning technologies make it possible to convert the electrodes to solid oxide fuel cells (SOFCs), which enables the enhancement of ionic conductivity and stability (Prajsner & Winiarz, 2024). Environmental and biomedical application is also developing at an alarming pace, where green synthesis routes synthesize graphene-silver nanocomposites with greater catalytic and antioxidant benefits (Rahim et al., 2024). Thermal transport engineering in composite materials, as well as the creation of ultralow thermal conductivity (TC) carbon nanocomposites indicate the key developments in thermoelectric energy conversion (Qiao et al., 2023; Rafique et al., 2023). Overall, these interdisciplinary studies highlight the centrality of CBNMs and innovative synthesis approaches to the overall solution of the world energy needs as well as the promotion of sustainable technologies.

H. Zhang (2024) proved that graphitic MoS_2/SnS/CoS heterostructures manufactured using the lattice-confinement method store a lot of sodium. Such ways to combine materials have been shown as a great strategy for the development of SIBs. Similarly, Yang and Liao (2024) proved that a self-supported network-like MoS_2/heteroatom-doped graphene has potential as a good anode for LIBs since it offers high capacity and very good cycle stability. The results point out that using graphene together with various materials greatly enhances SIB's efficiency. What's more, using several types of hard carbon is a main way to enhance SIB performance. According to Wang et al. (2023) hard carbon is great for energy storage owing to its good sodium storage properties and excellent performance in electrochemical reactions. It can also be shown that GBNs in sodium-ion batteries offer better energy levels, maintain stability, and charge/discharge quickly. The value of these materials will come from their mechanical sturdiness as well as from their ability to transfer ions for a green energy storage system, as these help the system earn economic benefits.

Figure 19 Graphene-based anodes in sodium-ion batteries.

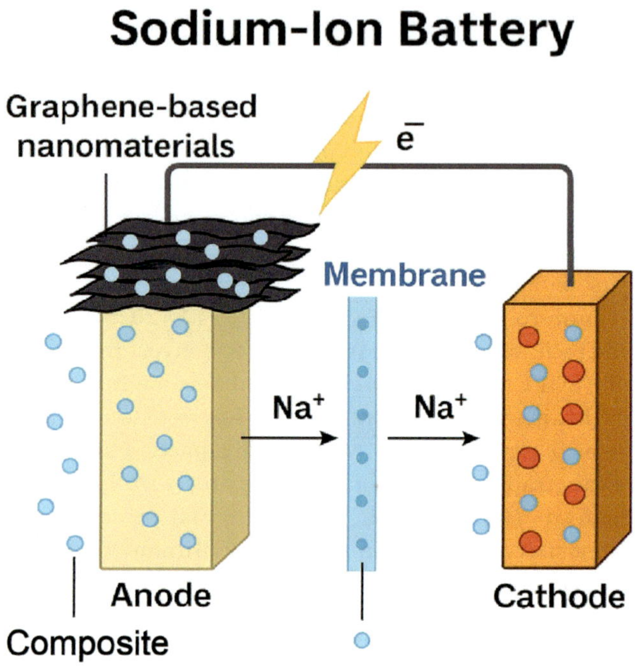

1.10.4 Solid-state batteries

Solid-state batteries (SSBs) can last longer than lithium-ion batteries, carry more power, and are safer too, so they can be a good replacement. Because batteries can be improved and balanced by graphene and its derivatives, SSBs have caught the attention of scientists. General graphene is a type of carbon that is perfectly formed and has excellent electro-mechanical properties in addition to a large specific surface. It could play a role in the enhancement of battery electrodes (Figure 20). Graphene-based nano-liquids address the issues that SSBs commonly have. Thanks to their contributions, the next-generation batteries can be more durable, cycle very quickly, and are safe. In general, researchers have found many methods to improve the safety of lithium-ion batteries. Despite how people might try to address this issue, the approaches can be sorted into two main types. One of the main strategies to notice early safety issues is by monitoring battery conditions as they happen. In a second strategy, efforts are put into bettering the battery's internal makeup by replacing or modifying the materials inside. Looking at the first strategy, many Battery

Thermal Management System (BTMS) have effectively prevented thermal runaway. In particular, when we consider BTMS hybrids, they offer excellent cooling abilities. Nevertheless, since these batteries are difficult and costly to work with, it is a big challenge for researchers to overcome. Fiber Bragg grating sensors make it possible to prevent security disasters by constantly monitoring lithium batteries' temperature, strain, and pressure. Such sensor devices can show an abnormality or heat irregularity. In the second strategy, the safety of lithium-ion batteries can be enhanced by using the best separator and electrolyte materials, blocking the creation of lithium fragments, and covering the surface of the cathode. As a result of implementing these strategies, lithium-ion batteries are much safer. Though it is on record that some lithium batteries are not safe, there are some limits to some research. Thus, extra attention in future should be given to the safety of lithium-ion batteries and materials to reach this improvement. Lithium-sulfur (Li-S) batteries use graphene as a conductor as well as a stopper to prevent polysulfides from getting out of the cathode and reducing effectiveness.

The electrical, mechanical, and chemical resourcefulness of graphene-based materials are shown to be significant in energy storage and conversion device transformation roles, thus continuing with the transformational role (Zhang, 2024). Execution of cycling activity under external pressure reveals considerable effects on the cycling performance of silicon-based lithium-ion batteries that the model and experimental validation produce mechanisms to improve the durability of electrodes and capacity retention (H. Zhang, 2024). Grand canonical Monte Carlo computational simulations allow understanding of how lithium-decorated polyhydroxyalkanoate (PHA)-graphene can be used to store hydrogen, allowing them to potentially be used in clean energy storage. The all-solid-state lithium-ion battery has also advanced in significance, including prelithiated graphite interfaces entrenched with a lithium fluoride interface to decrease stability and increase life span. Moreover, lattice-confinement synthesis of MoS_2/SnS/CoS heterostructures on graphene as substrate has been reported to increase sodium storage capacity to promise next-generation SIBs. Ionic conductivity and phase stability is important to solid-state battery performance and this is achieved through electrolyte materials that have been improved upon through Nb doping of $Li_7La_3Zr_2O_{12}$ solid electrolytes (D. Zhao et al., 2023). Carbon anodes made of pitch through N/S co-doping also add to the composition of high-performance SIB electrodes (J. Zhao et al., 2023). Hybrid energy storage is solving the issue of swift entry of renewable sources but also provides better flexibility and efficiency (Zhao et al., 2024). Machine learning methods foresee ultranarrow carbon nanowire bundles with large mechanical energy storage capacity, which demonstrates that there are new horizons of nanostructured energy apparatuses (J. Zhao et al., 2023). The evolution of printed SSBs and cathodes made of graphene-modified aluminum foil shows the increased strength and stability of the product on the mechanical level and advancing electrochemical properties, pointing toward the development of scalable and flexible energy storage devices (Zhou et al., 2023, 2024). All these multidisciplinary projects are aimed at opening the way to high-efficiency energy technologies and sustainable energy systems that can be obtained and developed on the basis of the state-of-the-art in energy materials.

High-performance energy storage devices and sustainable technologies are some of the recent elements of the growth of CBNMs. Porous carbons doped with nitrogen have impressive electrochemical performance, which promotes the higher conductivity and active sites quantity of attached supercapacitors (Zafar et al., 2023). Creative recycling of microplastic into graphene provides sustainable solutions to curbing plastic pollution, and at the same time creates precious nano materials (Zafar & Jacob, 2024). Graphene-like materials synthesized through deep eutectic solvents using biomass show sustainable options in the management of resources and energy utilizations (Zafar et al., 2024). The performance of silicon anode is enhanced greatly in lithium-ion batteries with the hybrid ionically covalently cross-linked binders, which solve the problems on volume expansion and cycling stability (Zeng et al., 2023a). Biologically derived hard carbon anodes are being commercialized as technologies develop to enhance the use of SIBs that allow cost and environmentally friendly energy sourced storage (Z. Zeng et al., 2023). Moreover, the synergistic effect of molybdenum dioxide-wrapped nitrogen-doped CNTs helps improve lithium storage capability in the binder-free anode, offering new possibilities of battery designing in the next generation (H. Zhang, 2024). All these multidisciplinary endeavors confirm the insertion point that engineered carbon nanomaterials would be applied in sustainable energy technologies and green remedies.

Nitrogen doped graphene (including N-Gr and N-rGO) has now caught the attention as possible new anodic materials to lead the development of future LIBs. Part of them may be used to make composites, and others might be used as medicinal products. All of these factors (nitrogen in the doped material, nitrogen doping layers, layers of N-Gr, oxalated groups, electroactive area, and nature of the material as 2D or 3D) can have an influence on the material's inner structure and chemical properties caused by N-Gr synthesis. All of these characteristics play an important role in how well the battery performs because they affect its number of active sites to lithium, lithium kinetics, travel within the system, resistance during charging and discharging, additional pseudo capacitive effects, and mechanical strength. Jauja-Ccana et al. (2023)

studied the different qualities of N-Gr materials and how they influence their performance as LIB anodes.

Figure 20 Graphene-enhanced solid-state battery.

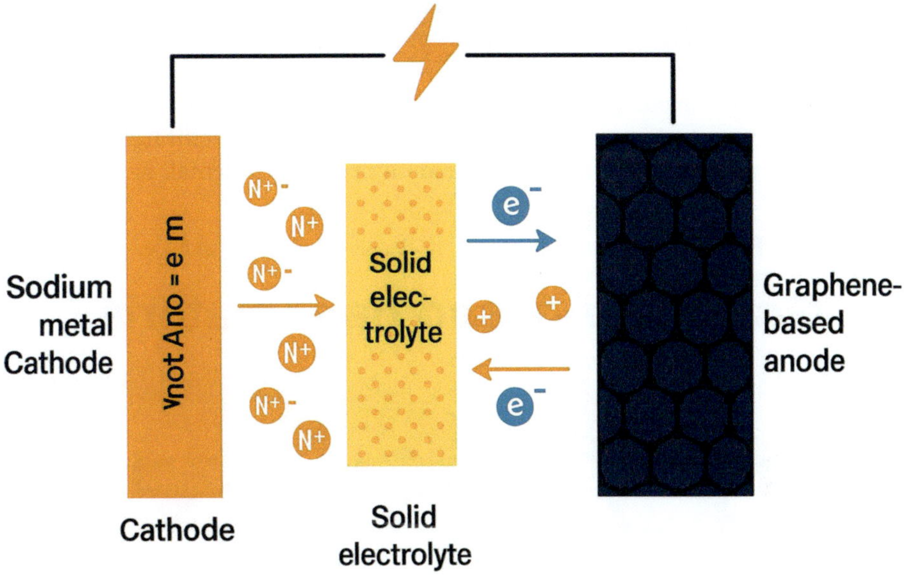

Graphene-like materials might improve the performance of SSBs. Due to its unmatched ability to allow quick movement of charge, the ability to be mechanically strong, and hold onto active materials, graphene is necessary for advanced energy storage devices. To aid development of SSBs, one should investigate how to enhance graphene-based composites and their communications with surrounding materials.

1.10.5 Hydrogen storage

Because graphene-based nanomaterials have a wide surface area, are durable and its electrical conductivity is versatile, this nanomaterial is studied widely for its hydrogen storage ability. As we can see in Figure 21, Pd and Ni catalysts are useful because they allow more hydrogen to be extracted. Graphene can be adjusted to achieve a much larger capacity for storing hydrogen. Key qualities that fuel cells and hydrogen fuel cells require to become new clean energy technologies are having a lot of storage space, a fast responsiveness, and the ability to work both ways, and this can be obtained by using graphene-based materials activated with transition metal catalysts.

Addition of 2.1 wt% palladium can improve the hydrogen uptake of the composite from zero to 0.42 wt% at 30°C and 40 bars compared to the unmodified graphene composites. Graphene materials covered with palladium could possibly be used for hydrogen storage since they are an environmentally friendly alternative to fossil fuels. The whole set of steps, from synthesizing to grafting, was accomplished according to green synthesis practices. The researchers reported that graphene decorated with Pd facilitates a greater number of hydrogens adapting to the amount present (Głowniak et al. 2024).

Figure 21 Graphene-based nanomaterials for hydrogen storage.

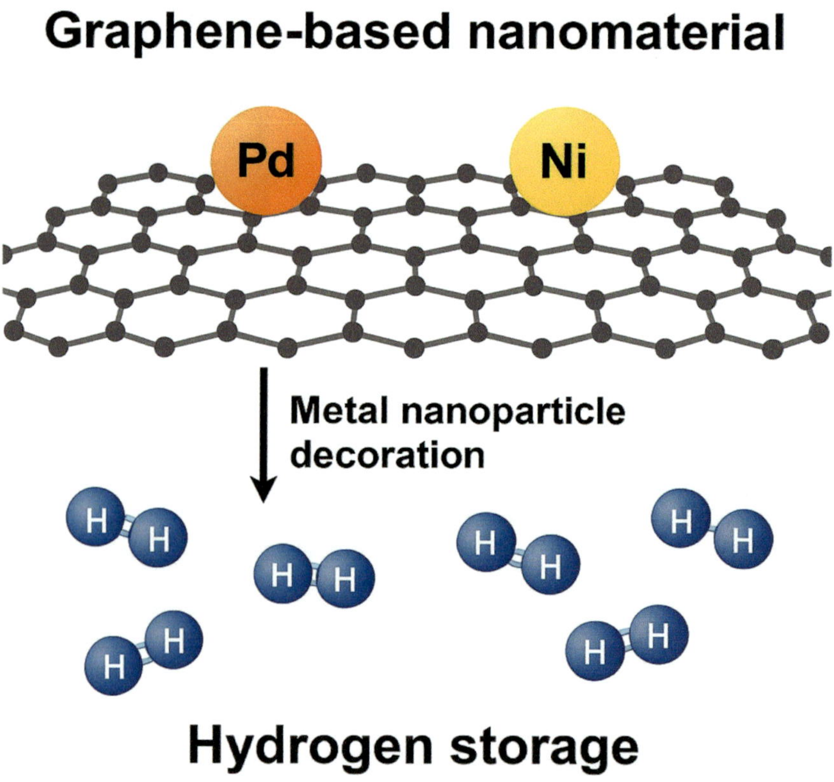

1.11 Energy Generation Applications

1.11.1 Fuel cells

Fuel cells involve nanomaterials made of graphene because they are needed for their exceptional conductivity and strong resistance to chemicals and mechanical stresses. Thanks to its special qualities, graphene can be applied in fuel cells as a catalyst, as a support, or a component of the membrane exchanging protons. One major result of using graphene in fuel cells is that it enhances the catalyst's efficiency. Indeed, one of the good things graphene can achieve in fuel cells is improvement of the catalyst. Platinum-sponsored catalytic materials are normally used in oxygen-reducing reactions in fuel cells, but platinum is rare and thus unaffordable, hence scientists are searching for a much more inexpensive and plentiful material to use. Because of this, graphene might serve as a supportive material to raise energy and stability during the use of these catalytic processes (J. Feng et al., 2024; Yarlagadda & Ramaswamy, 2025). The presence of graphene in nanocomposites is said to boost the overall functioning of fuel cells. For example, graphene-based nanocomposites make fuel cell components such as membranes stronger and improve their performance under high temperatures (Kausar et al., 2023).

Polymers of graphene were also designed for the purpose of making power devices better. In addition, due to grafting and handling of G/PEG nanocomposites, these adjustments have greatly helped boost graphene's charge storage and power generation abilities and resulted in equally high performance just like supercapacitors and related devices. So, graphene or graphene-derived nanofillers, methods for making graphene, and techniques for producing the nanocomposite that affect microstructural homogeneity, the connection between the matrix and the graphene, electronic current, specific capacitance, stored charge, enhanced charge storage ability, charge-discharge cycles, the devices' reliability, lifespan, stored energy density, and useful power of energy-related and similar systems are under discussion.

1.11.2 Solar cells

Starting from graphene-based materials, interests have arisen in solar energy conversion, the main focus being on making solar cells.

Figure 22 Comparative analysis of dye-sensitized solar cells performance with and without graphene.

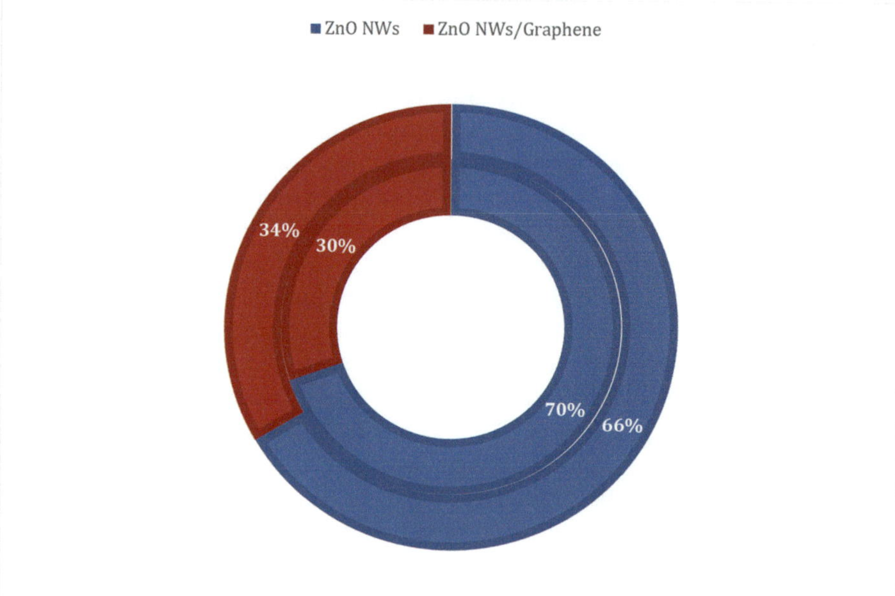

Due to its leading electricity conductivity (Table 4), expansive surface, and strong body, graphene is suitable for many types of photovoltaic devices. This section looks at how graphene helps make solar cells more effective and efficient, mainly by referring to DSSCs and other photovoltaic tools. Metal oxide composites made of graphene have the potential to boost how DSSCs operate. For instance, Fatima et al. argued that metal oxide mixed with graphene composites might be useful photoanode materials since they help with good charge transfer and light absorption necessary for DSSCs (Fatima et al., 2023). Development of DSSCs with pure ZnO nanowires and a ZnO nanowires/graphene bio-composite as electrodes was a major aspect of this work. Graphene was integrated into ZnO nanowires affordably inside a hydrothermal process and was then put in action as an electronic conductor. The current density-voltage (J-V) curve of the nanocomposite was checked using an irradiation of 100 mW per cm^{-2}.

The nanocomposites that were prepared were analyzed by using SEM, XRD, and J-V techniques. It is clear from SEM images that the ZnO NWs/graphene composite electrode is more porous than the electrode made with only ZnO NWs, allowing greater dye absorption and a bigger surface area. At the same time, graphene's low resistance and slight electron recombination loss allow better electronic flow to the collection electrode. DSSC2 shows excellent performance because the thickness of the ZnO NWs/graphene nanocomposite was 965.11 nm and it produces 19.50 $mA.cm^{-2}$ photocurrent and 9.829053% efficiency (Figure 22), which are different from measured results of DSSC1 without graphene. In the article of Ra et al. (2023), the authors provide an in-depth overview of the latest achievements in the study of heterostructures that can be used in optoelectronic devices. The paper addresses a broad range of low-dimensional materials such as 2D semiconductors, quantum dots or layered van der Waals heterostructures, with the emphasis on their distinctive electrical and optical properties. Strong focus has been on conduction mechanisms like tunneling, carrier confinement, and interlayer charge transfer, in search of high-performance photodetectors, LEDs, and solar cells. Ways of fabrication, interface engineering methods, and integration issues which influence the scalability and stability of the devices have also been investigated.

Table 4 Graphene performance values with integrations.

No.	Level	Jsc (mA/cm^2)	Effectivity (%)
1.	ZnO nanowires	14.30	6.20
2.	Graphene integrations	17.00	8.10
3.	ZnO NWs + graphene	19.50	9.83

1.11.3 Thermoelectric energy harvesting

Graphene and related materials have special properties and for this reason scientists have given much attention into using them in thermoelectric energy harvesting systems. Graphene has excellent strength and can hold or pass electricity. It holds huge potential to boost thermoelectric characteristics.

Thermoelectric devices made from graphene need low thermal but high electrical conductivity. Adding both MWCNTs and carbon black (CB) to FLG is suggested to make all-carbon composites with very low TC (κ), which may be useful for thermoelectric devices.

Due to the limitations of pure graphene, we should also consider mixing it with different materials to help with the thermoelectric energy conversion. Overall, future energy storage and conversion technologies may rely heavily on carbon and on CNTs above all. Researchers can use these special properties to innovate great supercapacitors and batteries.

1.11.4 Biofuel cells

Graphene used in biofuel cells addresses the problems of fast electron transfer, improved biocompatibility, and better performance altogether. One interesting aspect is that GO and rGO have good mechanical qualities, a big surface area, and high conductivity. It plays a big role in the popularity of EVs that charging an electric vehicle only takes around the same amount of time as stopping at a gas station to fill up a car with gasoline. The electrolyte is a main factor behind a battery's ability to charge fast since it determines the speed of ion movement and the interphase between the electrode and electrolyte at the two poles of the battery.

CHAPTER

2

Nanomaterials and Nanotechnology in Fuel Cells

2.1 Introduction to Carbon Nanostructured Materials in Fuel Cells

Table 5 explains how nanostructured carbon materials work in a fuel cell.

Table 5 Carbon nanostructured materials in fuel cells.

No.	Material	Role in Fuel Cell	Advantages	Common Fuel Cell Application
1	Graphene	Catalysts and extra support make electron movement more powerful	Big surface area and conducts electricity very well	Proton Exchange Membrane Fuel Cells (PEMFCs)
2	Carbon Nanotubes (CNTs)	Electrodes are made stronger and conductive by the addition of this material	1D structure makes it simple for electrons to travel and for gas to diffuse	Stacks, Direct Methanol Fuel Cells (DMFCs)
3	Carbon Nanofibers (CNFs)	Makes materials tougher and better at conducting any type of current	Eco-friendly materials, are easy to scale up, and tend to keep their properties well	PEMFCs or Solid Oxide Fuel Cells (SOFCs)
4	Mesoporous Carbon	There is more precious metal available on the surface for better use	This material can be made porous or compact and it holds charge very well	PEMFCs and Alkaline Fuel Cells (AFCs)
5	Activated Carbon	Substances used for the electrode to improve double-layer capacitance	Low price, strong surface area, and excellent adsorption abilities	Supercapacitors as well as Microbial Fuel Cells

DOI: 10.1201/9788743808213-2

Metals have stronger mechanics, remain stable chemically, and perform better in electrical and thermal functions than most materials. The unusual appearance of a nanotube makes it simple for electrons to move, so the structure is good for storing energy. Graphene has been used in composites, and it has been discovered to substantially boost the performance of supercapacitors because it helps store more charge and has greater energy density (Gopal et al., 2023; Ravisankar et al., 2024). Besides this, fuel cells and energy storage have looked into hollow carbon nanocages and CNTs. Strong mechanical and electrical properties of CNTs help improve many types of electrodes.

2.2 Nanoengineered Fuel Cells

One of the leading trends in energy conversion technologies are nanoengineered fuel cells because they are small and lightweight. Regarding efficiency, durability, and energy conversion, fuel cell performance is greatly enhanced by graphene's special properties (Figure 23). Choosing the best catalyst materials is a major question in nanoengineered fuel cells. Despite the fact that Pt is the top catalyst for Oxygen Reduction Reaction (ORR), its cost and rarity prevent it from being used everywhere; this is why it has attracted much research (P. Gao et al., 2023; Maulana et al., 2024). It appears that nanostructured options, particularly graphene and similar candidates, can strengthen both the stability and electrocatalytic performance of fuel cell electrodes.

Figure 23 Structure and function of nanoengineered fuel cell with graphene-based components.

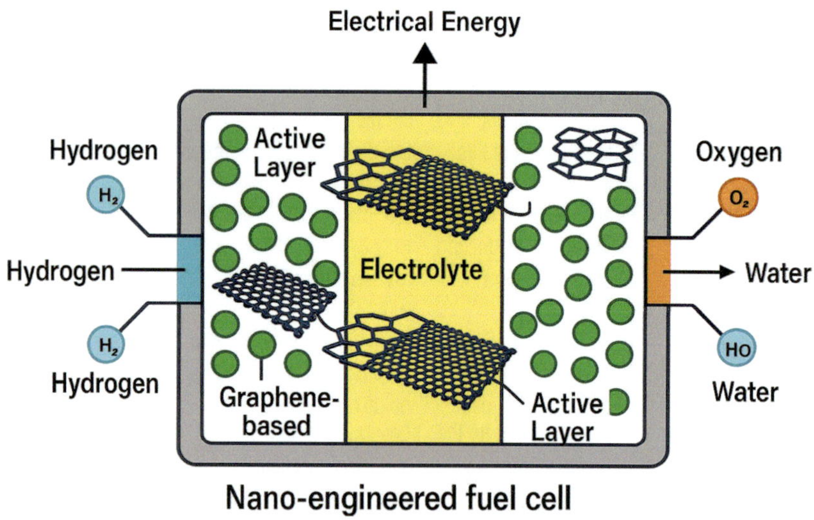

2.3 Emerging Nanomaterials and New Technologies

In the fields of medicine, farming, and treatment of the environment, new creative technologies and nanomaterials are now making a big impact in many areas. Thanks to their small size and the extra abilities they give, nanomaterials hold unique traits. This section will focus on how nanotechnology is utilized in different fields and supply evidence from existing research. Since graphene-based materials and new nanomaterials are used in electronics, medicine, and environmental protection, they have drawn lots of interest lately. Because of its mechanical, electrical, and thermal properties in the shape of graphene, a two-dimensional carbon atom lattice, this material's uses have become very valuable (Figure 24).

Figure 24 Applications of graphene-based nanomaterials across medicine, agriculture, and environmental science.

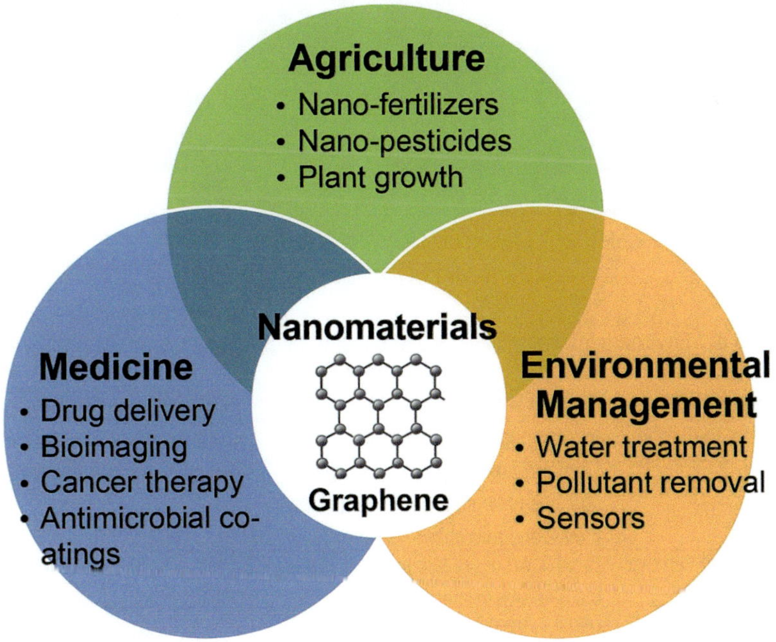

2.4 Environmental benefits

According to recent research, nanostructured graphene images can help make cancer detection more sensitive because less radiation is needed. Due to its high surface area, good strength, and the ability to conduct electricity, graphene is suitable for dealing with numerous environmental problems. They have unique features such as large surface area, good mechanical strength, and electricity conductivity, plus new applications using unique materials that are expected to help protect the environment (Figure 25 and Table 6).

Figure 25 Environmental benefits of graphene-based nanomaterials.

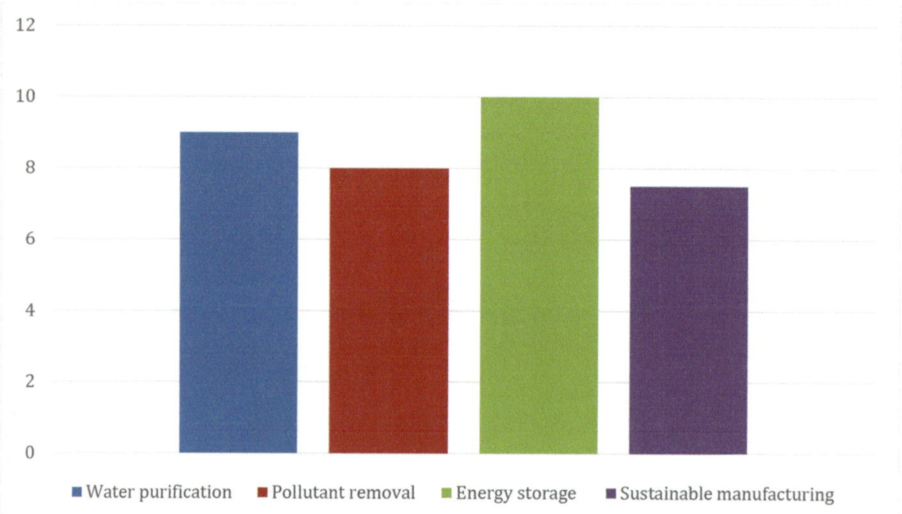

Table 6 Environmental applications and benefits of graphene-based nanomaterials.

No.	Application	Role of Graphene	Environmental Benefit
1	Water Purification	International Organization for Standardization collects and analyzes specimens for soil adsorption of heavy metals and microbes	Beneficiaries now find it easier to get clean water and good sanitation
2	Pollutant Removal	Breakdown of organic pollutants with the help of a catalyst	Industrial emitted pollution can be purified and made less toxic
3	Energy Storage	Metal is used to build electrodes in batteries and supercapacitors	Systems for renewable energy that work very efficiently

CHAPTER

3

Emerging Nanomaterials for Lithium-ion Batteries Energy Storage Devices

3.1 Introduction

New nanotech materials could enhance energy storage, battery endurance, and safety and these will encourage people to use it in LIBs. Thanks to their huge surface and unique electrochemical features, nanoparticles are very significant for cathode and anode uses. One of the top advantages of nanomaterials is their help in improving the exchange of ions and electrons in LIBs (Table 7).

Table 7 Nanomaterials in lithium-ion battery components.

No.	Component	Nanomaterial Used	Function	Benefit
1	Anode	Graphene, silicon nanoparticles	Improve the ability of lithium-ion to spread and increase the flow of electrons	Phones now hold more power and charge faster
2	Cathode	$LiFePO_4$ with rGO, $LiCoO_2$ with CNTs	Build better charge, increase cycling stamina, and boost speed of charging.	Being able to store more power in the batteries for extended use
3	Electrolyte	Nanostructured solid electrolytes	Improve the ability to carry ions and tolerate high temperatures	An improvement in how safe and comfortable it is at different operational temperatures
4	Separator	Nano-coated polymer separators	Stop the growth of dendrites and strengthen the product's structure	Greater safety and an increased service period

DOI: 10.1201/9788743808213-3

Graphene and CNTs included at the nanoscale make it quicker for lithium-ions to travel and diffuse in the battery. The process of charging and discharging becomes quicker and the cycle of the battery may change (Liu, 2023a; Sama, 2023). Furthermore, nanomaterials secure and extend the life of LIBs. After adding metal nanoparticles to lithium batteries, it is documented that they provide much-needed safety and high energy, along with solutions to overheating and losing their energy capacity (Z. Chen, 2024; Ren, 2023; Tian, 2024). Results also suggest that adding ZnO-carbon nanocomposites to LIBs can help enhance their energy density as well as the ability to maintain stability throughout a large number of cycles (Febrian et al., 2024; Savitha et al., 2024). All things taken into account, applying different nanomaterials can let EV batteries do more with better performance, safety, reliability, and capacity. Even though more research is needed, searching for brand new nanomaterials and their uses could play a big part in finding new ways to store energy in various areas.

3.2 Synthesis and Applications

The efficiency of LIBs has risen with nanomaterials, which has impacted the way we store energy. In this section, the key points about nanomaterials in LIBs and how they are manufactured and put into use is explored.

3.3 Role of Nanomaterials in Lithium-ion Batteries

Modern energy storage devices for portable gadgets and electric vehicles depend largely on lithium-ion batteries. How the batteries perform depends a lot on the materials in their electrodes.

Nanomaterials have been documented to increase a number of electrochemical properties. Researchers are applying nanotechnology to enhance both the anodes and cathodes in batteries, making them hold more charge, function longer, and operate more efficiently. Nanomaterials including CNTs and graphene are under study by scientists to check if they can improve the conductivity and size of the surface areas in the electrode material (Figure 26).

Lithium ions can easily move on the large and electricity-conductive CNT surface and the electrodes for batteries might show better electrochemical performance (Ma, 2024). If silicon is taken as an example (Geng, 2024), nanostructured materials might assist in handling the problem of volume increase as batteries are charged and discharged over many cycles.

Figure 26 Nanomaterial-enhanced electrodes in lithium-ion batteries for improved energy performance.

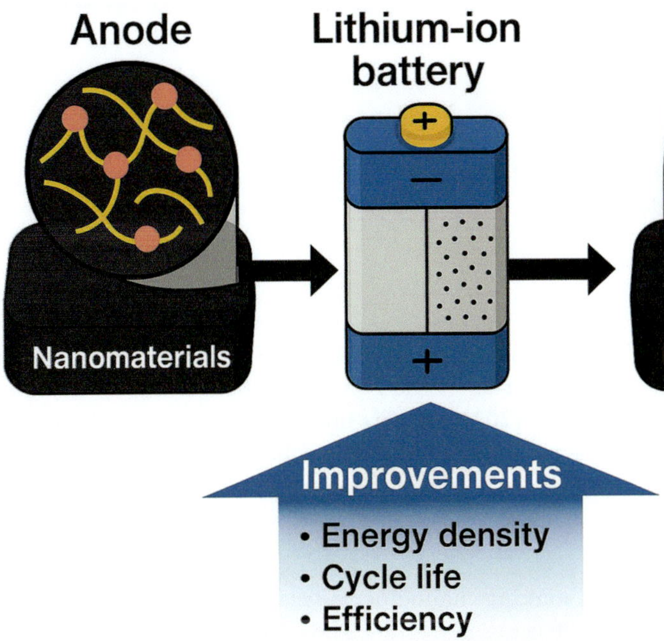

Experts continue discovering new information about nanostructured materials and nanocomposites. They aim to build batteries with improved densities, long-lasting cycles, and improved safety to handle modern technologies' requirements.

3.3.1 How do nanomaterials improve LIBs?

They make the electrode more conductive, speed up the diffusion of ions, give responses a bigger surface area, and make cycling more stable and long-lasting.

3.4 Key Nanomaterials for LIBs

Nanomaterials included in LIBs are chosen especially because they can improve the performance of batteries by increasing energy density, steady cycling, and the overall efficiency. Metal oxides, different advanced composites, carbons, silicon-based materials, and other materials are the main nanomaterials used in LIBs.

3.4.1 Carbon nanomaterials

They are made of carbon, graphene and CNTs have remarkable surface area and conduct electricity very well.

3.4.2 Silicon nanomaterials

Due to its powerful capacity (4.200 mAh/g), silicon may also be an excellent material for LIB anodes.

3.4.3 Metal oxides

The focus on Li_4Ti5O_{12} (LTO) as an anode material comes from its good stability and exceptional safety. The batteries of the future will be made with LTO, as described by Julien and Mauger (2024), as this nano-structured material helps lithium batteries perform better in many ways (Sundhar et al., 2023).

3.4.4 Porous carbon materials

Researchers indicate that making nanoelectrodes porous would increase the storage of lithium-ions. Porous materials allow the lithium-ion to work efficiently by increasing the area available to them and letting the electrolyte go further inside for the best cycle of charging and discharging (Chen et al., 2024). To improve their strength and electrical conductivity, there may be optimal designs for porous structures.

3.4.5 Composite materials

It has been noticed that mixing several types of nanomaterials in hybrid types improves LIBs performance. When metallic nanoparticles interact with carbon dioxide in batteries, there might be a better effect on both the conductivity and capacity (Tian, 2024). Using this strategy, it is possible to design things that can cope with mechanical stresses due to charging and discharging lithium-ions.

3.4.6 Layered two-dimensional materials

New research has shown that good LIB materials are now believed to be two-dimensional (2D) and include layers of oxides and transition metal materials, with an extra atom known as chalcogen, because of their excellent conductivity and lithium-ion-absorbing properties.

3.5 Anode Materials

When silicon is used as a nanostructure, it can be considered as nanoparticles or coated with carbon. Although its theoretical capacity is almost 4200 mAh/g, using nanostructured silicon as an anode for lithium-ion batteries is limited because it has major volume changes during the reaction and this may bring about fracture and capacity losses. Silicon nanoparticles and carbon coatings have been offered as some of the possible solutions to these challenges.

3.5.1 Coatings of carbon

One answer to keep silicon anodes from swelling is covering them with a layer of carbon. The shifting in volume from cycling gives the carbon material a chance to bear stress.

3.5.2 Silicon nanoparticles

Using silicon nanoparticles could fix one of the issues related to volume expansion. Researchers have pointed out that a porous design of silicon/carbon composites allows the electrode to stay firm even when it expands (Yang & Liao, 2024).

3.5.3 Hybrid structures

Hybrid technologies promising modern constructions that use silicon combined with other substances are in abundance.

3.5.4 Advanced binders

To maintain silicon's structure in the anode, it is very important to select the most suitable binder materials. Researchers have conducted tests on binders that are able to resists the pressure caused by volume changes. As an example, Zeng et al. (2023a) recommend using special binders that mix covalent bonds and ion exchange to withstand changes in size during lithiation. For this reason, silicon anodes have better long-term stability. Using a self-healing polymer binder can result in better binding and stability of the silicon on the anode that stays stable as the volume changes (Bai et al., 2024).

3.5.5 Porous silicon structures

Researchers working in this field have tested numerous processes to stop the silicon from growing in volume. Porous silicon helps the anode survive for a longer cycle, showing an overall better performance as it allows the volume of the anode to change without causing trouble. To run properly, the battery requires easy access by the electrolyte and broad movement of lithium-ions inside the electrode.

Key factors in design of anode materials include:

Electrical conductivity enhancement: It was demonstrated by Kubarkov et al. that replacing classic CB with CNTs or graphene enables a greater decrease of conductive material with the same or higher conductivity, since they achieved percolation thresholds at lower values (Kubarkov et al., 2023).

Stopping particles from sticking together: The use of nanomaterials in LIBs is challenging because they sometimes cluster and reduce the batteries' efficiency. CNTs and graphene are expected to be a part of the solution. Due to its large area and strong interactions, graphene or CNT-templated composites keep the electrode materials distributed and stop them from gathering during the cycling process (Zhao et al., 2023c). It is very important for the electrochemical performance to be maintained by this stabilization capability.

Structures made up of parts: Experiments have proven that including graphene and CNTs in LIB materials leads to an enhanced performance of LIB batteries. Making use of CNTs and graphene forms a three-dimensional architecture, which is stated by L. Lin et al. (2023), to fully benefit from the materials found in lithium-sulfur batteries. It will make electron transfer happen faster. These systems are able to extract the best qualities of the materials and get better conductivity and stability.

Anodes that work really well: Many studies focus on using graphene and CNTs as anodes in lithium-ion batteries. According to Chen (2025), a little amount of graphene can let anodes gain more energy and prevent voltage polarization problems that might come up when using a larger amount of graphene.

Mechanochemical procedures: Producing graphene from mechanical processes can be expected to become more effective in lithium-ion application (G. Lin et al., 2024). This introduces new ways to boost the efficiency of batteries.

Challenges and future prospectives of anode materials include:

Structural stabilization via composite formation: Building composite materials is one way to help long-term stability in transition metal oxides.

Carbon coatings: Another approach for stabilizing transition metal oxides is carbon coating. Furthermore, adding a conductive carbon layer may raise the mechanical support and electrical conductivity of the underlying oxide material.

Doping strategies: This processing of the material may improve its structural stability, making it less resistant and creating a more stable cycle.

Nanostructuring: It might be possible to develop these nanostructured transition metal oxides and have them work better by making the surface area larger and letting lithium-ions go to the materials faster.

More advanced ways to make things: Setting up metal oxides in new ways might support their stability. With better control over the manufacturing process, one can ensure that the shell structures stay stable in the battery cycle. Using sol-gel, hydrothermal, and electrospinning, it is now possible to get clearly defined nanostructures with better mechanical and electrochemical properties from transition metal oxides.

Hybrid materials: Transition oxides mixed with graphene or CNTs may help form hybrids that optimize the qualities of both materials. For example, introducing carbon to Fe_3O_4 leads to better conductivity as well as better cycling performance.

3.6 Cathode Materials

Nanoparticles of lithium iron phosphate ($LiFePO_4$) have a very long useful life and do not get damaged. For this reason, $LiFePO_4$ nanoparticles are appealing for use in the cathode materials in LIBs. This section unites recent information on improving $LiFePO_4$ nanoparticles' physical features and what they can be used for.

3.6.1 Thermal stability

One of the best things about $LiFePO_4$ is that it can handle heat quite well.

3.6.2 Cycle life

The excellent cycling property of the $LiFePO_4$ nanoparticles is what allows lithium-ion batteries to last a long time.

3.6.3 Good for the environment

$LiFePO_4$ is known for being good for the environment since it doesn't include heavy metals like cobalt or nickel, which are often present in other cathode materials.

3.6.4 Cost-effectiveness

$LiFePO_4$ also stands out because it is very affordable. It is not costly to obtain the required chemicals, so the technology for making it on a larger scale might be improved.

3.6.5 Doping and composite strategies

Various elements have been added to $LiFePO_4$ to try to increase its levels of conductivity and electrochemical performance. The material's performance could improve if it undergoes doping, as it changes both the electrical conductivity and lithium-ion diffusion in $LiFePO_4$ (Jiao et al., 2023).

3.7 Nanostructuring Modifications

3.7.1 Better capacity retention by nanosizing

Nanosizing layered oxides greatly improves their electrochemical performance by making the surface area bigger and the routes for lithium-ions to diffuse shorter.

3.7.2 Better kinetics and rate capability

The nanoscale size of layered oxides makes electrochemical processes happen quicker. Moghaddam et al. (2023) observed that when manganese oxides are made into nanosized layers, they speed up the diffusion of lithium-ions, which is important for high-rate applications. The larger surface area not only gives lithium-ions more places to interact, but it also makes the electrode material more conductive, which speeds up the battery's rate capability.

3.7.3 Stability of structure

Nanosizing also helps keep layered oxides stable throughout cycling.

3.7.4 Composite approaches

Making composite materials that mix nanosized layered oxides with conductive additions like CNTs or graphene may make performance even better.

3.7.5 Doping strategies

It may be helpful to include additional materials in stacked oxides in order to increase their electrochemical abilities.

3.7.6 Synthesis techniques

Layered nano oxides can be made with controlled properties using the sol-gel and hydrothermal methods. This way, nanoparticles can be made so that they are all similar and have better electrochemical properties. Doing the synthesis of $LiCoO_2$ by hydrothermal way rather than the conventional method results in higher capacity and more stable.

3.8 Key Elements to a Better Design

3.8.1 Efficient lithium-ion diffusion

The three-dimensional lithium-ion diffusion routes in the spinel structure of $LiMn_2O_4$ make lithium insertion and extraction faster.

3.8.2 Better cycling performance

It is proven that processing $LiMn_2O_4$ into nanoparticles helps it keep its stability and high capacity for longer. With finer particles, the lithium-ions have a shorter journey for each cycle, which media accelerates diffusion and makes the battery's overall electrochemical performance better. Small nanosized $LiMn_2O_4$ manages changes in volume more effectively and as a result, it can cycle more than the larger size particles.

3.8.3 Composite materials

If $LiMn_2O_4$ is combined with conductive materials (primarily CNTs or graphene) the battery performance can get better.

3.8.4 Ways to make a synthesis

Methods such as sol-gel and hydrothermal synthesis are used now to make $LiMn_2O_4$ in nano form with a carefully controlled shape and composition. As a result of these strategies, nanoparticles are made evenly and perform better in electrochemistry.

3.8.5 Benefits for the environment and the economy

$LiMn_2O_4$ serves as a good cathode material because it costs less and the raw material is safe for the economy and the environment.

3.9 Electrolytes and Separators

3.9.1 Better ionic conductivity

The ionic conductivity of LLZO is especially high and may reach more than 10^{-3} S/cm at room temperature. Because of this, lithium battery charging and discharging are done much quicker.

3.9.2 Better safety

Using LLZO and similar electrolytes is safer than using the conventional liquid types. Because LLZO does not catch fire and is stable, the possibilities of overheating and leaks in its battery cells are reduced.

3.9.3 Structural stability

LLZO is very structurally stable, it is able to maintain good performance throughout many cycles. LLZO stays stable no matter how the battery is used, it is assured that the solid electrolyte will keep its conductivity and won't degrade with time.

3.9.4 Nanocomposite methods

Introducing specialized carbon matter or other nanostructures might make the electrolyte more ready to conduct electricity and stronger.

3.9.5 Synthesis techniques

LLZO has been produced with a desired form and composition by applying advanced methods such as sol-gel procedures and ball milling. As a result of these techniques, nanoparticles can be made that are all the same size and more conductive to ions.

3.9.6 Future prospects

Gaining more knowledge about LLZO and other solid-state electrolytes is helping scientists come up with better battery technology. Currently, scientists are trying to mix LLZO solid-state electrolytes with different materials to improve their performance. To sum up, using electrolytes made with LLZO is safer and their conductivity of ions is better.

3.10 Nanostructures Polymer Electrolytes

3.10.1 Better ionic conductivity

A big advantage of nanostructured polymer electrolytes is that they could have greater ionic conductivity than those that are not nano.

3.10.2 Flexibility and mechanical qualities

Since lithium-ion batteries are bent to fit in various devices, having high mechanical strength is important in polymer electrolytes.

3.10.3 Better safety

Solid and nano forms of polymers have higher safety than liquid electrolytes.

3.10.4 Composite methods

Incorporating several nanostructured components in composite polymer electrolytes could make their performance better.

3.10.5 Synthesis techniques

With electrospinning and solution casting methods, polymer electrolytes can be made with specific forms and levels of ingredients. Such techniques can make sure that you obtain nanofibers or nanoparticles that help the electrolyte's electrochemical properties.

3.10.6 Future prospects

With more demand for flexible and fast-working batteries, nanostructured polymer electrolytes are likely to be important in developing the next set of lithium-ion batteries. Nanostructured polymer electrolytes are better than other types since they have better ion transfer, are flexible, safe, and work more efficiently in lithium-ion batteries.

3.11 Nanofiber Separators

3.11.1 Stopping dendrite formation

Preventing the formation of needle-like dendrites that can appear on the lithium metal anode as it is charged is one of the main tasks of nanofiber separators.

3.11.2 Better ionic conductivity

To improve the battery system's conductivity by adding nanofiber separators could help. Due to their big surface and pores, electrolytes can soak into nanofibers and ions can pass through them with ease. Polyimide electrospun nanofiber separators play an important role by absorbing enough electrolyte to keep the battery's performance high as it runs, according to X. Song et al. (2024).

3.11.3 Mechanical stability and flexibility

The flexible separators make space for any changes in volume that occur during plating and unplaiting the lithium-ions.

3.11.4 Composite materials

Adding more components to these materials could make them perform better than before. In order to allow lithium-ions to flow more swiftly and to make nanofiber separators more conductive, Wang et al. worked on adding conductive parts to them (C. Wang et al., 2023). Combining the matrix of the polymer with the best components is done in these composite materials to boost their electrochemical performance.

3.11.5 Methods of synthesis

Thanks to electrospinning and other similar techniques, it is possible to make nanofiber separators that have required shapes and characteristics. One can use this to produce the same type of nanofibers that could serve different needs.

3.11.6 Future prospects

Nanofiber separators research regularly leads to improvements in battery technology. At the moment, scientists are studying how to incorporate nanofiber separators with conductive polymers and metal-organic frameworks (MOFs). Thanks to hybrid technologies, lithium-ion batteries may become safer and perform better, which would allow their use in various other areas (Xia et al., 2023). For a summary of applications of nanomaterials in LIBs, see Table 8.

Table 8 Applications of nanomaterials.

Purpose	Adaptation
Batteries with a lot of energy	Silicon nanostructures and layered oxide cathodes
Fast-charging LIBs	Nanosized electrodes make it easier for ions to move around
Batteries that can be worn and are flexible	Gel electrolytes and nanofiber-based electrodes
Next-generation solid-state batteries	Nanocomposite solid electrolytes make them safer
LIBs that are good for the environment	Bio-derived nanocarbon materials have less of an influence on the environment

CHAPTER

4

Nanoengineered Materials and Fourth-generation Solar Cells

The use of nanotechnology increases the performance, adjustability, and affordability of photovoltaic systems, which is a major development in the solar industry. With the help of nanoscale materials like quantum dots, nanowires, and hybrid substances, nano solar cells are different from silicon solar cells. Being able to adjust nano solar cells for each purpose is an important benefit. They could be designed to be light and flexible so that it becomes possible to place them on portable devices, on building projects, and clothes. In addition, new technologies can be applied to create nano solar cells that can be produced using long manufacturing rolls, which may lower the cost. While lowering manufacturing expenses, scientists investigate nano solar cells in order to increase how well they generate power, how reliable they are, and their potential for production (Figure 27).

4.1 Introduction to Nano Solar Cells

In the development of nano solar cells, nanotechnology is making a major change to solar energy. With the unique characteristics of nanomaterials, these new solar cells become more efficient, less costly, and more sustainable in making electricity. The use of solar energy in alleviating the world energy problem depends on nano-technologies. Quantum dots and perovskites are examples of nanostructured materials that can boost light absorption and movement of charges in solar cells (a remarkable progress is now being seen in the field of nano cells).

DOI: 10.1201/9788743808213-4

Figure 27 Nano solar cell.

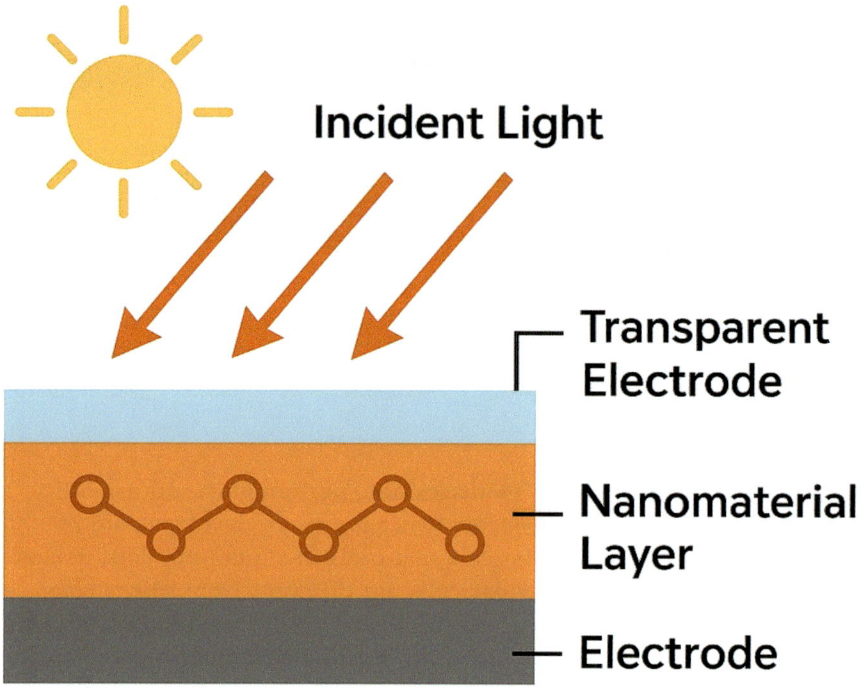

Advances in nanotechnology are changing the process of using solar energy. Thanks to nanotechnology, we can make solar cells work more efficiently, make progress in energy storage, and implement sustainable solutions so energy can be used in a better way. However, to solve today's issues and maximize the abilities of nano solar cells, more research and development are needed.

4.2 Significance of Fourth-generation Solar Cells

It is especially in PSCs that fourth-generation solar cells surpass earlier versions in terms of efficiency, cost-effectiveness, and flexibility. Fourth-generation solar cells are different from the earlier generations because they combine organic and inorganic materials. One of the greatest improvements in fourth-generation solar cells is the rise in power conversion efficiency (see Figure 28).

Figure 28 Fourth-generation solar cell.

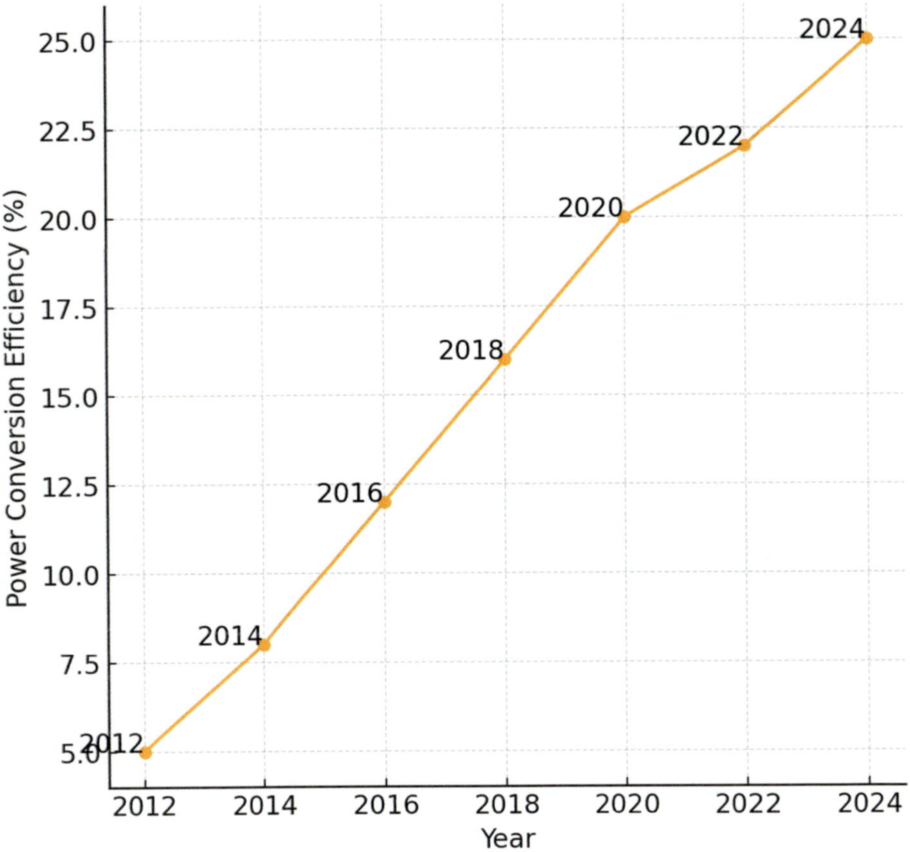

In 2009, PSCs worked at an efficiency of 3.8%, and now their performance is standing above 26% (Mai et al., 2024). In addition, PSCs are a suitable solution for large solar energy projects because they can be manufactured from cheap substances and produced on a large scale (Srivishnu et al., 2023). Being flexible, PSCs can be fitted into portable electronic devices and BIPV (building-integrated photovoltaics). Being flexible allows new designs that fit the modern demand for architecture and electronics, which opens up more sales opportunities for solar energy market.

4.3 Advantages of Nanomaterials

4.3.1 Improved performance

The physical and chemical traits of nanomaterials are quite unlike their larger-sized equivalents. Catalysis and drug delivery are made better with nanoparticles because their large surface-area-to-volume ratio increases how easily they combine with other materials. Quantum dots and perovskites in solar cells made with nanomaterials increase the amount of light that is absorbed and improves charge transfer, which raises the efficiency of these cells (Nazir et al., 2023; Simpa et al., 2024).

4.3.2 Cost-effectiveness

Traditional ways of producing materials are usually more complicated and costly than new nanoparticle techniques. Using nanomaterials, DSSCs can be built under normal atmosphere, which greatly decreases production costs (Hasan & Susan, 2024; Rajendhiran et al., 2023). Furthermore, choosing green approaches with plant extracts helps to decrease the negative impact on the environment and the expense of producing nanomaterials (Devade et al., 2024).

4.3.3 Versatility and customization

Nanomaterials can be custom designed to display traits that are precise for their intended roles.

4.3.4 Environmental benefits

The introduction of better energy sources and the reduction of hazardous waste are possible with the help of nanomaterials.

4.4 Design of Next-generation Solar Cells

The new solar cells are looking to overcome the issues found in silicon-based solar cells by improving efficiency, minimizing costs, and increasing flexibility and sustainability. With better manufacturing, nanotechnology, and material research, scientists are producing solar cells that require less space and can be used more widely as demand for renewable energy rises (Figure 29). Designers of next-generation solar cells want to achieve better efficiency, affordable prices, and fit them into a variety of devices. Because of the range of materials, efficiency, flexibility, eco-friendliness, and smart technology, solar energy is now available to many instead of a niche option. As they are studied and improved, solar cells are predicted to play an important role in meeting the world's energy needs and building a safer, more energy-saving world.

Figure 29 Design of next-generation solar cells.

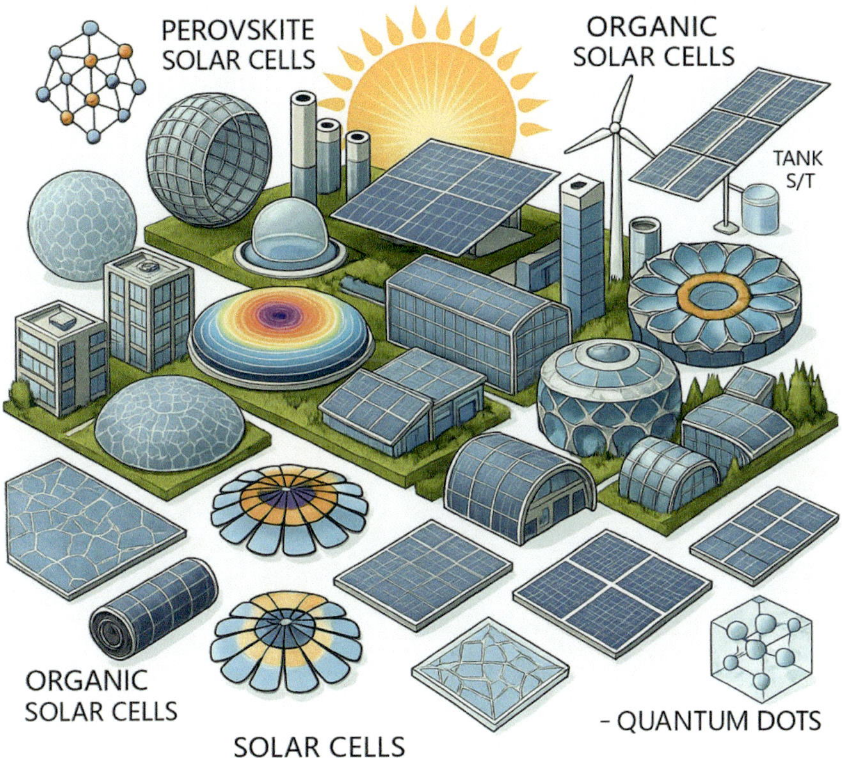

Benefits such as performance, being cost-effective, adaptability and sustainability have increased over the years from 2012 to 2025. The trend in developing new solar cells is clear from Figure 29, especially as it explains how advances in materials, new manufacturing methods, and being environmentally friendly are emphasized. Developing new energy targets for the world needs a big improvement in performance and sustainability, as the statistics state. Moreover, to make solar energy the main energy source, people seek options that are cheaper and have many uses; as a result, various options for solar farms, small devices, and places to use them are now shared. By 2025, it is likely that solar energy will be proven to be very important in renewable energy projects across the globe.

References and Further Reading

Abaszade, R.G. (2022). Synthesis and analysis of flakes graphene oxide. *Journal of Optoelectronic and Biomedical Materials,* 14(3), 107–114. https://doi.org/10.15251/JOBM.2022.143.107

Abaszade, R.G., Kapush, O.A., Mamedova, S.A., Nabiyev, A.M., Melikova, S.Z., & Budzulyak, S.I. (2020a). Gadolinium doping influence on the properties of carbon nanotubes. *Physics and Chemistry of Solid State,* 21(3), 404–408. https://doi.org/10.15330/pcss.21.3.404–408

Abaszade, R.G., Kapush, O.A., & Nabiyev, A.M. (2020b). Properties of carbon nanotubes doped with gadolinium. *Journal of Optoelectronic and Biomedical Materials,* 12(3), 61–65. https://www.chalcogen.ro/61_AbaszadeRG.pdf

Abaszade, R.G., Mammadova, S.A., Kapush, O.A., Agayev, F.G., Nabiev, A.M., Mammadova, M.M., Budzulyak, S.I., & Kotsyubynsky, V.O. (2021). Synthesis and characterization of graphene oxide flakes for transparent thin films. *Physics and Chemistry of Solid State,* 23(3), 595–601.

Abaszade, R.G., Mammadov, A.G., Bayramov, I.Y., Khanmamadova, E.A., Kotsyubynsky, V.O., & Kapush, O.A. (2022). Structural and electrical properties of sulfur-doped graphene oxide/graphite oxide composite. *Physics and Chemistry of Solid State,* 25(2), 256–260.

Abaszade, R.G., Aliyev, E.M., Babanli, M.B., Kotsyubynsky, V.O., Zapukhlyak, R.I., Mammadov, A.G., Budak, H.F., Kasapoglu, A.E., Gür, E., Margitych, T.O., & Stetsenko, M.O. (2023a). Investigation of thermal properties of carbon nanotubes and carboxyl group-functionalized carbon nanotubes. *Physics and Chemistry of Solid State,* 24(3), 530–535. https://doi.org/10.15330/pcss.24.3.530–535

Abaszade, R.G., Tatardar, F.N., Mamadov A.G., & Khanmamadova, E.A. (2023b). *Microelectronics and Nanoelectronics*, ISBN 978-9952-39-042-1 (in Azeri).

Abaszade, R.G., Mammadov, A.G., Khanmamedova, E.A., Aliyev, F.G., Kotsyubynsky, V.O., Gür, E., Soltabayev, B.D., Margitych, T.O., Stetsenko, M.O., Singh, A., & Arya, S. (2024). Photoconductivity of functionalized carbon nanotubes, *Digest Journal of Nanomaterials and Biostructures*, 19(2), 837–843. https://doi.org/10.15251/DJNB.2024.192.837

Adeola, A., Duarte, M., & Naccache, R. (2023). Microwave-assisted synthesis of carbon-based nanomaterials from biobased resources for water treatment applications: emerging trends and prospects. *Frontiers in Carbon*, 2. https://doi.org/10.3389/frcrb.2023.1220021

Adhi, T., Sumampouw, G., Pramudita, D., Munandari, A., Kurnia, I., Mohtar, W., et al. (2024). Sustainable management of water resources for drinking water supply by exploring nanotechnology. *Water*, 16(13), 1896. https://doi.org/10.3390/w16131896

Adhikari, C., Ghosh, S., & Das, P. (2023). Graphene's role in the clean-up act: progress in industrial wastewater treatment. *Asian Journal of Chemistry*, 36(1), 1–17. https://doi.org/10.14233/ajchem.2024.30639

Aftab, S., Saeed, M., Hussain, S., Kabir, F., Aslam, M., Rajpar, A., et al. (2023). Perovskite photovoltaics: navigating stability challenges for enhanced efficiency. *Solar RRL*, 8(2). https://doi.org/10.1002/solr.202300803

Ahmad, I. & Kausar, A. (2023). Graphene and nanocomposites—imprints on environmentally sustainable production and applications based on ecological aspects. *Characterization and Application of Nanomaterials*, 6(2), 4226. https://doi.org/10.24294/can.v6i2.4226

Ahmad, I., Kausar, A., Zhao, T., Aldaghri, O., Ibnaouf, K., & Eisa, M. (2023). Graphene nanocomposites as innovative materials for energy storage and conversion—design and headways. *International Journal of Molecular Sciences*, 24(14), 11593. https://doi.org/10.3390/ijms241411593

Ahn, Y., Li, T., Huang, S., Ding, Y., Hwang, S., Wang, W., et al. (2023). Mixed-phase niobium oxide as a durable and ultra-fast charging anode for high-power lithium-ion batteries. *Advanced Functional Materials*, 34(8). https://doi.org/10.1002/adfm.202310853

Ai, S., Wu, X., Wang, J., Li, X., Hao, X., & Meng, Y. (2024). Research progress on solid-state electrolytes in solid-state lithium batteries: classification, ionic conductive mechanism, interfacial challenges. *Nanomaterials*, 14(22), 1773. https://doi.org/10.3390/nano14221773

Ajdari, F., Asghari, P., Aghdam, A., Abbasi, F., Rao, R., Abbasi, A., et al. (2024). Silicon solid state battery: the solid-state compatibility, particle size, and carbon compositing for high energy density. *Advanced Functional Materials*, 34(30). https://doi.org/10.1002/adfm.202314822

Al-Dayel, I., Nadeem, M., & Khan, M. (2024). On comparative analysis of graph entropies of symmetrical carbon nanotube Y-junctions. *Physica Scripta*, 99(5), 055237. https://doi.org/10.1088/1402-4896/ad37ae

Al-Hamdani, Y., Zen, A., & Alfè, D. (2023). Unraveling H_2 chemisorption and physisorption on metal decorated graphene using quantum Monte Carlo. *The Journal of Chemical Physics*, 159(20). https://doi.org/10.1063/5.0174232

Al-Hamdani, Y., Zen, A., Michaelides, A., & Alfè, D. (2023). Mechanisms of adsorbing hydrogen gas on metal decorated graphene. *Physical Review Materials,* 7(3). https://doi.org/10.1103/physrevmaterials.7.035402

Alathlawi, H. & Hassan, K. (2024). Review—recent advancements in graphene-based electrodes for lithium-ion batteries. *ECS Journal of Solid State Science and Technology,* 13(1), 011002. https://doi.org/10.1149/2162-8777/ad15a7

Ali, H. (2023). Preparation and characterization of ZnO NWs/graphene nanocomposite as an effective photoanode electrode for improving the performance of the dye-sensitized solar cells. *University of Thi-Qar Journal of Science,* 10(1), 11–15. https://doi.org/10.32792/utq/utjsci/v10i1.915

Ali, S., Khan, M., & Irfan, M. (2024). Thermal performance enhancement of lithium-ion batteries using phase change material and fin geometry modification. *World Electric Vehicle Journal,* 15(2), 42. https://doi.org/10.3390/wevj15020042

Ali, S., Shahzad, T., & Hussain, M. (2024). Enhanced electrochemical performance of $FeCO_2S_4$@g-C_3N_4 composites for advanced energy storage applications. *Ann Civ Eng Manag,* 1(1), 01–08. https://doi.org/10.33140/acem.01.01.07

Anagbonu, P., Ghali, M., & Allam, A. (2023). Low-temperature green synthesis of few-layered graphene sheets from pomegranate peels for supercapacitor applications. *Scientific Reports,* 13(1). https://doi.org/10.1038/s41598-023-42029-w

Anjum, F., Shaban, M., Ismail, M., Gul, S., Bakhsh, E., Khan, M., et al. (2023). Novel synthesis of CuO/GO nanocomposites and their photocatalytic potential in the degradation of hazardous industrial effluents. *ACS Omega,* 8(20), 17667–17681. https://doi.org/10.1021/acsomega.3c00129

Apfelbeck, F., Heger, J., Zheng, T., Guan, T., Schwartzkopf, M., Roth, S., et al. (2024). Influence of the polymer binder composition on the charge transfer resistance, morphology, and crystallinity of $LiFePO_4$ electrodes revealed by electrochemical impedance spectroscopy and grazing incidence small- and wide-angle X-ray scattering. *Small Science.* https://doi.org/10.1002/smsc.202400154

Arifin, N., Noor, E., Mohamad, F., & Mohamad, N. (2024). The impact of spinning speed on n-TiO_2/ZnO bilayer thin film fabricated through sol–gel spin-coating method. *Coatings,* 14(1), 73. https://doi.org/10.3390/coatings14010073

Arshad, M., Mbianda, X., Ali, I., Wanbing, G., Kamal, T., Kauhaniemi, K., et al. (2023). Advances and perspectives on solid oxide fuel cells: from nanotechnology to power electronics devices. *Energy Technology,* 11(12). https://doi.org/10.1002/ente.202300452

Arslan, S., Yalçın, Z., Da , M., & Akın, M. (2024). The role of carbon in lead-acid batteries: applications, challenges, and future opportunities. *Uluslararası*

Batı Karadeniz Mühendislik Ve Fen Bilimleri Dergisi, 6(2), 53–79. https://doi.org/10.55440/umufed.1554371

Atchudan, R., Alagan, M., Kishore, S., Perumal, S., Wadaan, M., Baabbad, A., et al. (2023). Recent advanced synthesis strategies for the nanomaterial-modified proton exchange membrane in fuel cells. *Membranes,* 13(6), 590. https://doi.org/10.3390/membranes13060590

Báez, D. (2023). Graphene-based nanomaterials for photothermal therapy in cancer treatment. *Pharmaceutics,* 15(9), 2286. https://doi.org/10.3390/pharmaceutics15092286

Bagade, S., Patel, S., Malik, M., & Patel, P. (2023). Recent advancements in applications of graphene to attain next-level solar cells. *C – Journal of Carbon Research,* 9(3), 70. https://doi.org/10.3390/c9030070

Bagherdoost, M. & Askarzadeh, A. (2023). Techno-economic optimization of a hybrid system composed of pumped hydro storage, photovoltaic module, and diesel generator in a multi-objective framework. *Environmental Progress & Sustainable Energy,* 43(2). https://doi.org/10.1002/ep.14279

Bai, H., Zhou, Z., Yang, B., Liu, W., Liu, L., & Wen, J. (2024). A self-healing polymer binder of silicon-based anode with enhanced lithium-ion transport performance. *Journal of the Electrochemical Society,* 171(5), 050513. https://doi.org/10.1149/1945-7111/ad4398

Bardwell, L., Rahbari, A., & Pye, J. (2024). Hydrogen production via off-sun solar-thermal supercritical water gasification and membrane reforming of piggery waste. *SolarPACES Conf Proc,* 1. https://doi.org/10.52825/solarpaces.v1i.748

Baruah, A., Newar, R., Das, S., Kalita, N., Nath, M., Ghosh, P., et al. (2024). Biomedical applications of graphene-based nanomaterials: recent progress, challenges, and prospects in highly sensitive biosensors. *Discover Nano,* 19(1). https://doi.org/10.1186/s11671-024-04032-6

Benettayeb, A., Seihoub, F., Pal, P., Ghosh, S., Usman, M., Chia, C., et al. (2023). Chitosan nanoparticles as potential nano-sorbent for removal of toxic environmental pollutants. *Nanomaterials,* 13(3), 447. https://doi.org/10.3390/nano13030447

Bhardwaj, A., Okoroanyanwu, U., Pagaduan, J., Fan, W., & Watkins, J. (2024). Large-area fabrication of porous graphene networks on carbon fabric via millisecond photothermal processing of polyaniline for supercapacitors. *Small,* 20(33). https://doi.org/10.1002/smll.202402049

Bi, J., Du, Z., Sun, J., Liu, Y., Wang, K., Du, H., et al. (2023). On the road to the frontiers of lithium-ion batteries: a review and outlook of graphene anodes. *Advanced Materials,* 35(16). https://doi.org/10.1002/adma.202210734

Biehler, E., Quach, Q., & Abdel-Fattah, T. (2023). Screening study of different carbon-based materials for hydrogen storage. *ECS Journal of Solid State*

Science and Technology, 12(8), 081002. https://doi.org/10.1149/2162-8777/acec12

Bist, A., Pant, B., Ojha, G., Acharya, J., Park, M., & Saud, P. (2023). Novel materials in perovskite solar cells: efficiency, stability, and future perspectives. *Nanomaterials,* 13(11), 1724. https://doi.org/10.3390/nano13111724

Boateng, E., Zalm, J., Burns, N., Chow, D., Kycia, S., & Chen, A. (2023). Tunable electrochemical hydrogen uptake and release of nitrogen-doped reduced graphene oxide nanosheets decorated with Pd nanoparticles. *ACS Applied Nano Materials,* 6(18), 17311–17323. https://doi.org/10.1021/acsanm.3c03763

Bongu, C., Tasleem, S., Krishnan, M., & Alsharaeh, E. (2024). Graphene-based 2D materials for rechargeable batteries and hydrogen production and storage: a critical review. *Sustainable Energy & Fuels.* https://doi.org/10.1039/d4se00497c

Brewster, C., Terry, L., Doan, H., Rochat, S., & Ting, V. (2023). Hydrogen sorption on microporous carbon/sulfur nanocomposite systems. *Energy Advances,* 2(3), 398–409. https://doi.org/10.1039/d2ya00242f

Bristy, H., Rono, N., Shams, D., Bashar, M., Sultana, M., & Sikder, A. (2024). Characterization of flower-based organic dyes and titanium dioxide thin film for dye-sensitized solar cell. *IOP Conference Series: Materials Science and Engineering,* 1305(1), 012036. https://doi.org/10.1088/1757-899X/1305/1/012036

Byun, J., Kim, W., Lee, M., Yu, J., Choi, J., Yun, J., et al. (2024). Osteoporosis failure of aluminum current collector induced crosstalk degradation at the imide-type lithium salt comprised practical-level lithium-ion batteries. *Journal of the Electrochemical Society,* 171(6), 060536. https://doi.org/10.1149/1945-7111/ad5a3e

Cabeza, L., Martínez, F., Borri, E., Ushak, S., & Prieto, C. (2024). Thermal energy storage using phase change materials in high-temperature industrial applications: multi-criteria selection of the adequate material. *Materials,* 17(8), 1878. https://doi.org/10.3390/ma17081878

Cai, K., Xiang, C., Wang, X., Zhang, X., Zhang, D., Zhen, Z., et al. (2024). Construction of artificial SEI on graphite electrode through electropolymerization of p sulfonated polyallyl phenyl ether. *Journal of the Electrochemical Society,* 171(3), 030505. https://doi.org/10.1149/1945-7111/ad2db2

Cao, X., Gao, P., Kautz, D., Matthews, B., Xu, Y., Han, K., et al. (2023). Designing electrolytes with controlled solvation structure for fast-charging lithium-ion batteries. *Advanced Energy Materials,* 13(35). https://doi.org/10.1002/aenm.202301199

Chai, J., Song, L., Li, Z., Peng, Z., & Yao, X. (2024). Lithium spreading layer consisting of nickel particles enables stable cycling of aluminum anode in all-solid-state battery. *Battery Energy,* 3(5). https://doi.org/10.1002/bte2.20240004

Chalmpes, N., Tantis, I., Alsmaeil, A., Aldakkan, B., Dimitrakou, A., Karakassides, A., et al. (2025). Elevating waste biomass: supercapacitor electrode materials derived from spent coffee grounds. *Energy & Fuels,* 39(2), 1305–1315. https://doi.org/10.1021/acs.energyfuels.4c05250

Chatterjee, P. & Dhibar, S. (2023). Nanomaterial marvels: pioneering applications and cutting-edge advancements in drug delivery. *Nano and Medical Materials,* 220. https://doi.org/10.59400/nmm.v3i2.220

Chaudhuri, A., Arya, V., Bakli, C., Bera, C., & Kumar, N. (2023). Significantly reduced thermal conductivity and enhanced thermoelectric performance of twisted bilayer graphene. *Journal of Applied Physics,* 134(4). https://doi.org/10.1063/5.0153052

Chen, E. (2024). Strategies for improving the safety of lithium-ion batteries. *Highlights in Science Engineering and Technology,* 83, 174–180. https://doi.org/10.54097/y8s18r94

Chen, F., Zhang, X., Guan, X., Gao, S., Hao, J., Li, L., et al. (2023). Nitrogen-doped or boron-doped twin T-graphene as advanced and reversible hydrogen storage media. *Applied Surface Science,* 622, 156895. https://doi.org/10.1016/j.apsusc.2023.156895

Chen, H. (2024). Application of porous carbon nanoelectrode materials for lithium-ion batteries. *Journal of Physics Conference Series,* 2798(1), 012036. https://doi.org/10.1088/1742-6596/2798/1/012036

Chen, H., Zheng, Y., Li, J., Li, L., & Wang, X. (2023). AI for nanomaterials development in clean energy and carbon capture, utilization and storage (CCUS). *ACS Nano,* 17(11), 9763–9792. https://doi.org/10.1021/acsnano.3c01062

Chen, J. & Lü, X. (2023). Effect of different materials on the performance of flexible organic solar cells. *Applied and Computational Engineering,* 7(1), 663–669. https://doi.org/10.54254/2755-2721/7/20230542

Chen, N., Zhang, G., Chen, H., & Yue, H. (2023). Conductive carbon-wrapped fluorinated hard carbon composite as high-performance cathode for primary lithium batteries. *Coatings,* 13(5), 812. https://doi.org/10.3390/coatings13050812

Chen, R., Zhou, Y., & Li, X. (2023). Fe3C/nanocarbon-enabled lithium dendrite mitigation in lithium–sulfur batteries. *Small,* 20(17). https://doi.org/10.1002/smll.202308261

Chen, X. (2025). Research progresses on nano-structured silicon-based materials as anode for lithium-ion batteries. *Materials,* 18(4), 830. https://doi.org/10.3390/ma18040830

Chen, Y., Jiang, B., Zhao, Y., Liu, H., & Ma, T. (2023). Diatomite and glucose bioresources jointly synthesizing anode/cathode materials for lithium-ion batteries. *Coatings,* 13(1), 146. https://doi.org/10.3390/coatings13010146

Chen, Y., Xia, G., Jin, C., Wang, Y., Yan, Y., Chen, Y., et al. (2023). Hydrogen storage properties of economical graphene materials modified by non-precious metal nickel and low-content palladium. *Inorganics,* 11(6), 251. https://doi.org/10.3390/inorganics11060251

Chen, Z., Hung, F., & Chang, K. (2024). The solid-state battery applicational technology: material characteristics and charge–discharge mechanisms of iron chloride electrodes. *Energy Storage,* 6(6). https://doi.org/10.1002/est2.70044

Chikkatti, B., Sajjan, A., Banapurmath, N., & Ayachit, N. (2023). Graphene-doped hydrogels promoting ionic conductivity in gel-valve-regulated lead acid batteries. *Langmuir,* 39(48), 17232–17239. https://doi.org/10.1021/acs.langmuir.3c02285

Choi, C., Yun, T., & Hwang, B. (2023). Dispersion stability of carbon nanotubes and their impact on energy storage devices. *Inorganics,* 11(10), 383. https://doi.org/10.3390/inorganics11100383

Choi, J., Lee, H., Lee, M., Lee, S., Kang, J., Suh, J., et al. (2024). Multi-interface strategy for electrode tailoring toward fast-charging lithium-ion batteries. *Advanced Functional Materials,* 34(33). https://doi.org/10.1002/adfm.202400414

Chu, Y., Wang, P., Ding, Y., Lin, J., Zhu, X., Zhao, S., et al. (2023). High-capacity Sb/Fe2S3 sodium-ion battery anodes fabricated by a one-step redox reaction, followed by ball milling with graphite. *ACS Applied Materials & Interfaces,* 15(20), 24354–24365. https://doi.org/10.1021/acsami.3c01223

Dai, X., Ma, X., Hu, D., Duan, J., & Chen, H. (2024). An overview of the R&D of flywheel energy storage technologies in China. *Energies,* 17(22), 5531. https://doi.org/10.3390/en17225531

Dai, Z., Ma, Y., Li, Y., Zhang, Y., Wang, G., Luo, Y., et al. (2024). Nanostructured carbon materials derived from lignin via flame-induced oxidation for supercapacitors. *ACS Applied Nano Materials,* 7(23), 26743–26755. https://doi.org/10.1021/acsanm.4c04517

Daniels, K., Harrington, J., Wiseall, A., Shoemark-Banks, E., Hough, E., Wallis, H., et al. (2023). Battery earth: using the subsurface at Boulby Underground Laboratory to investigate energy storage technologies. *Frontiers in Physics,* 11. https://doi.org/10.3389/fphy.2023.1249458

Davies, E. (2024). Innovations in energy through nanomaterials enhancing solid-state battery safety and efficacy. *Highlights in Science Engineering and Technology,* 116, 250–256. https://doi.org/10.54097/fjrk7k69

De, A., Bhattcharjee, J., Chowdhury, S., & Roy, S. (2023). A comprehensive review on third-generation photovoltaic technologies. *Journal of Chemical Engineering Research Updates,* 10, 1–17. https://doi.org/10.15377/2409-983x.2023.10.1

Devade, K., Singh, P., Kumar, S., Kumar, H., Prasad, B., Rao, A., et al. (2024). Green nanotechnology-based sustainable energy solutions and environmental impacts. *E3S Web of Conferences,* 511, 01031. https://doi.org/10.1051/e3sconf/202451101031

Dhiman, N., Sharma, V., & Ghosh, S. (2023). Perspective on biomass-based cotton-derived nanocarbon for multifunctional energy storage and harvesting applications. *ACS Applied Electronic Materials,* 5(4), 1970–1991. https://doi.org/10.1021/acsaelm.3c00022

Ding, C., Shen, Z., Zhu, Y., & Cheng, Y. (2023). Insights into the modification of carbonous felt as an electrode for vanadium redox flow batteries. *Materials,* 16(10), 3811. https://doi.org/10.3390/ma16103811

Ding, H. & Zhang, X. (2023). Sodium intercalation in nitrogen-doped graphene-based anode: a first-principles study. *Crystals,* 13(7), 1011. https://doi.org/10.3390/cryst13071011

Ding, L. & Pei, J. (2024). On improved active disturbance rejection controller design for a pumped storage hydro unit. *Journal of Physics Conference Series,* 2707(1), 012070. https://doi.org/10.1088/1742-6596/2707/1/012070

Ding, X., Zhou, Q., Li, X., & Xiong, X. (2024). Fast-charging anodes for lithium-ion batteries: progress and challenges. *Chemical Communications,* 60(18), 2472–2488. https://doi.org/10.1039/d4cc00110a

Ding, Z., Li, S., Jiang, Y., Wang, D., & Yuan, M. (2023a). Open-circuit voltage loss in perovskite quantum dot solar cells. *Nanoscale,* 15(8), 3713–3729. https://doi.org/10.1039/d2nr06976h

Ding, Z., Tang, Y., Ortmann, T., Eckhardt, J., Dai, Y., Rohnke, M., et al. (2023b). The impact of microstructure on filament growth at the sodium metal anode in all-solid-state sodium batteries. *Advanced Energy Materials,* 13(48). https://doi.org/10.1002/aenm.202302322

Dragoman, M., Dinescu, A., Aldrigo, M., Dragoman, D., Mohebbi, E., Pavoni, E., et al. (2024). Graphene monolayer nanomesh structures and their applications in electromagnetic energy harvesting for solving the matching conundrum of rectennas. *Nanomaterials,* 14(19), 1542. https://doi.org/10.3390/nano14191542

Du, M., Advincula, P., Ding, X., Tour, J., & Xiang, C. (2023). Coal-based carbon nanomaterials: en route to clean coal conversion toward net zero CO_2. *Advanced Materials,* 35(25). https://doi.org/10.1002/adma.202300129

Du, W., Wang, G., Li, Y., & Yu, Y. (2024). Development of fullerene acceptors and the application of non-fullerene acceptors in organic solar cells. *Frontiers in Physics,* 12. https://doi.org/10.3389/fphy.2024.1378909

Du, Y., Wang, M., Ye, X., Liu, B., Han, L., Jafri, H., et al. (2023). Advances in the field of graphene-based composites for energy–storage applications. *Crystals,* 13(6), 912. https://doi.org/10.3390/cryst13060912

Duan, H., Meng, D., & Yuan, S. (2023). Solution combustion synthesis of high-performance nano-$LiFePO_4$/C cathode material from cost-effective mixed fuels. *Materials,* 16(22), 7155. https://doi.org/10.3390/ma16227155

Duan, P., Zhang, R., Wu, Z., Yang, Z., Li, H., Zhong, Y., et al. (2023). Design and construction of porous silicon materials as stable anodes for lithium-ion batteries. *Industrial & Engineering Chemistry Research,* 62(18), 6995–7006. https://doi.org/10.1021/acs.iecr.3c00084

Duraia, E., Opoku, M., & Beall, G. (2024). Efficient eco-friendly synthesis of carbon nanotubes over graphite nanosheets from yellow corn: a one-step green approach. *Scientific Reports,* 14(1). https://doi.org/10.1038/s41598-024-65893-6

Düsterhaupt, S., Černíková, M., & Hyblerová, Š. (2024). Economic evaluation of kinetic energy storage systems as key technology of reliable power grids. *PLOS One,* 19(10), e0311160. https://doi.org/10.1371/journal.pone.0311160

Dutra, A., Rudman, G., Johnston, K., & Dawson, J. (2023). Defect chemistry and ion transport in low-dimensional-networked Li-rich anti-perovskites as solid electrolytes for solid-state batteries. *Energy Advances,* 2(5), 653–666. https://doi.org/10.1039/d3ya00075c

Eatmon, Y., Stiles, J., Hayashi, S., Rupp, M., & Arnold, C. (2023). Air stabilization of $Li_7P_3S_{11}$ solid-state electrolytes through laser-based processing. *Nanomaterials,* 13(15), 2210. https://doi.org/10.3390/nano13152210

Eidhagen, J., Larsson, A., Preobrajenski, A., Delblanc, A., Lundgren, E., & Pan, J. (2023). Synchrotron XPS and electrochemical study of aging effect on passive film of Ni alloys. *Journal of the Electrochemical Society,* 170(2), 021506. https://doi.org/10.1149/1945-7111/acba4b

El-Kalliny, A., Abdel-Wahed, M., El-Zahhar, A., Hamza, I., & Gad-Allah, T. (2023). Nanomaterials: a review of emerging contaminants with potential health or environmental impact. *Discover Nano,* 18(1). https://doi.org/10.1186/s11671-023-03787-8

Elghamry, I., Gouda, M., & Al-Faiyz, Y. (2023). Synthesis of chemically modified acid-functionalized multiwall carbon nanotubes with benzimidazole for removal of lead and cadmium ions from wastewater. *Polymers,* 15(6), 1421. https://doi.org/10.3390/polym15061421

Erduran, V., Bayat, R., Işık, İ., Bayazıt, T., & Şen, F. (2023). Production of sustainable low-layer graphene by green synthesis at room conditions for platinum-based direct methanol fuel cell. *ACS Engineering Au,* 3(6), 537–545. https://doi.org/10.1021/acsengineeringau.3c00040

Esmer, S. & Bekiroğlu, E. (2022). Design of PMaSynRM for flywheel energy storage system in smart grids. *International Journal of Smart Grid,* (6)4. https://doi.org/10.20508/ijsmartgrid.v6i4.255.g254

Fan, S., Cui, S., Zhang, J., Rong, J., Wang, W., Xing, X., et al. (2023). Two birds with one stone: micro/nanostructured SiO_xC_y composites for stable Li-ion and Li metal anodes. *Small,* 19(48). https://doi.org/10.1002/smll.202304290

Fan, W., Gou, J., Huang, Y., She, K., Yu, M., & Zhang, Z. (2024). "Peapod-like" fiber network: a universal strategy for composite solid electrolytes to inhibit lithium dendrite growth in solid-state lithium metal batteries. *Nano Letters,* 24(29), 9050–9057. https://doi.org/10.1021/acs.nanolett.4c02274

Farid, M., Sabidi, S., Ruozhu, W., Lease, J., Maeda, T., Tsubota, T., et al. (2023). Fabrication of graphene-based nanofiltration membrane from fractionated bamboo lignin for dye removal and desalination. *ACS Applied Engineering Materials,* 1(10), 2791–2808. https://doi.org/10.1021/acsaenm.3c00498

Fatima, K., Pandith, A., Manzoor, T., & Qureashi, A. (2023). DFT studies on a metal oxide@graphene-decorated D–ϖ_1–ϖ_2–A novel multi-junction light-harvesting system for efficient dye-sensitized solar cell applications. *ACS Omega,* 8(9), 8865–8875. https://doi.org/10.1021/acsomega.3c00333

Febrian, R., Septiani, N., Iqbal, M., Nursam, N., & Muliani, L. (2024). CV performance of ZnO-carbon nanocomposite based on ZIF-8 as lithium-ion battery anode. *Journal of Physics Conference Series,* 2705(1), 012013. https://doi.org/10.1088/1742-6596/2705/1/012013

Feng, A., Zhu, X., Chen, Y., Liu, P., Han, F., Zu, Y., et al. (2024). Functional biomass-derived materials for the development of sustainable batteries. *ChemElectroChem,* 11(13). https://doi.org/10.1002/celc.202400086

Feng, J., Xu, S., Xia, C., Song, N., Dong, H., Yu, L., et al. (2024). Density functional theory study of the relationship between n-dopants and vacancy defects on graphene quantum dots for oxygen reduction electrocatalysis. *ACS Applied Nano Materials,* 7(18), 21578–21589. https://doi.org/10.1021/acsanm.4c03480

Feng, S., Li, K., Hu, P., Cai, C., Liu, J., Li, X., et al. (2023). Solvent-free synthesis of hollow carbon nanostructures for efficient sodium storage. *ACS Nano,* 17(22), 23152–23159. https://doi.org/10.1021/acsnano.3c09328

Feng, X., Ning, Y., Wu, Z., Li, Z., Xu, C., Li, G., et al. (2023). Defect-enriched graphene nanoribbons tune the adsorption behavior of the mediator to boost the lactate/oxygen biofuel cell. *Nanomaterials,* 13(6), 1089. https://doi.org/10.3390/nano13061089

Ferdous, A., Shah, S., Shah, S., Johan, B., Bari, M., & Aziz, M. (2024). Transforming waste into wealth: advanced carbon-based electrodes derived from refinery and coal by-products for next-generation energy storage. *Molecules,* 29(9), 2081. https://doi.org/10.3390/molecules29092081

Festus-Ikhuoria, I., Obiuto, N., Olajiga, O., & Adebayo, R. (2024). The role of nanomaterials in energy storage: a comparative review of USA and

African development. *World Journal of Advanced Research and Reviews,* 21(3), 2073–2083. https://doi.org/10.30574/wjarr.2024.21.3.0925

Ficca, V., Sbroscia, M., Stellino, E., Rago, I., Goto, F., Majumdar, I., et al. (2024). Non-chemical route to PGM-free via N^+ ion implantation in vertically aligned carbon nanotubes. *Advanced Functional Materials,* 35(4). https://doi.org/10.1002/adfm.202413308

Fonseca, A. & Nagy, J.B. (2001). "Carbon nanotubes formation in the arc discharge process," in *Carbon Filaments and Nanotubes: Common Origins, Differing Applications?* Biro, L.P., Bernardo, C.A., Betts, G.G., Lambin, Ph., Eds., Dordrecht: Kluwer Academic Publishers, 75–84. https://doi.org/10.1007/978-94-010-0777-1_5

Fonseca, A., Scalon, L., Vale, B., Guaita, M., Bettini, J., Brandão, Z., et al. (2024). In situ PL tracking of halide exchange at 3D/QD heterojunction perovskite solar cells. *ACS Energy Letters,* 9(6), 3177–3186. https://doi.org/10.1021/acsenergylett.4c01268

Freitas, D., Filho, C., Loguercio, L., Santa-Cruz, L., Maciel, L., Kohlrausch, E., et al. (2023). Advances in carbon materials applied to carbon-based perovskite solar cells. *Energy Technology,* 11(4). https://doi.org/10.1002/ente.202200676

Freitas, M., Silva, L., Rodrigues, P., & Silva, J. (2024). Sustainable technological applications of green carbon materials. *Sustainable Chemistry,* 5(2), 81–97. https://doi.org/10.3390/suschem5020007

Fu, L., Zheng, Y., Li, X., Liu, X., Lin, C., & Karimi-Maleh, H. (2023). Strategies and applications of graphene and its derivatives-based electrochemical sensors in cancer diagnosis. *Molecules,* 28(18), 6719. https://doi.org/10.3390/molecules28186719

Fu, M., Ye, Y., Niu, Y., Guo, S., Wang, Z., & Liu, X. (2024). Graphene-based tunable dual-frequency terahertz sensor. *Nanomaterials,* 14(4), 378. https://doi.org/10.3390/nano14040378

Gamo, H., Maeda, Y., Kiyobayashi, T., Siroma, Z., & Sano, H. (2024). Elucidating the mechanism of microscopic conduction in cathode composites for all-solid-state batteries through scanning spreading resistance microscopy. *Journal of Materials Chemistry A.* https://doi.org/10.1039/d4ta01634c

Gan, H., Sun, J., Ye, X., Qiu, P., Li, L., & Zhu, W (2023). PEO/PVDF–HFP/LLZTO composite solid polymer electrolyte for high-performance all-solid-state lithium metal batteries. *The Journal of Physical Chemistry C, 127(43),* 21015–21021. https://doi.org/10.1021/acs.jpcc.3c06275

Gao, P., Kautz, D., Matthews, B., Xu, Y., Han, K., Cao, X., et al. (2023). Designing electrolytes with controlled solvation structure for fast-charging lithium-ion batteries. *Advanced Energy Materials,* 13(35). https://doi.org/10.1002/aenm.202301199

Gao, Y., Li, Z., Wang, P., Cui, W., Wang, X., Yang, Y., et al. (2024). Experimentally validated design principles of heteroatom-doped-graphene-supported calcium single-atom materials for non-dissociative chemisorption solid-state hydrogen storage. *Nature Communications,* 15(1). https://doi.org/10.1038/s41467-024-45082-9

Gao, Y., Wang, Y., Cao, Y., Hu, Y., Wang, G., Han, M., et al. (2024). Hybrid powder-based additive manufacturing of laser-induced graphene 3D architectures with tunable porous microstructures from waste sources of black liquor and white pollution. *Advanced Sustainable Systems,* 8(12). https://doi.org/10.1002/adsu.202400565

Gao, Z., Chen, Z., Zhan, X., Zhou, L., Xie, Y., Yang, X., et al. (2023). Pt nanoparticles supported on iron and nitrogen-doped holey graphene for boosting ORR performance. *ACS Applied Nano Materials,* 6(12), 10521–10530. https://doi.org/10.1021/acsanm.3c01502

Garg, R., Balakrishnan, G., Rashid, R., Gershanok, S., Roman, D., Wang, Y., et al. (2023). Graphene and poly(3,4-ethylenedioxythiophene)–polystyrene sulfonate hybrid nanostructures for input/output bioelectronics. *ACS Applied Nano Materials,* 6(10), 8495–8505. https://doi.org/10.1021/acsanm.3c00849

Geng, Y. (2024). Application of nanotechnology in the negative electrode of Si-based lithium-ion batteries. *Highlights in Science Engineering and Technology,* 116, 208–213. https://doi.org/10.54097/15fnr841

Ghosh, S. & Saha, B. (2023). Estimation of energy harvesting by thermoelectric cement composites with nanostructured graphene and metallic oxides. *Journal of Composites Science,* 7(5), 207. https://doi.org/10.3390/jcs7050207

Ghotia, S., Rimza, T., Singh, S., Dwivedi, N., Srivastava, A., & Kumar, P. (2024). Hetero-atom doped graphene for marvellous hydrogen storage: unveiling recent advances and future pathways. *Journal of Materials Chemistry A,* 12(21), 12325–12357. https://doi.org/10.1039/d4ta00717d

Głowniak, S., Szczęśniak, B., Choma, J., & Jaroniec, M. (2024). Greener approach towards synthesis of palladium-decorated graphene derivatives for hydrogen adsorption. *ChemPhysChem,* 25(5). https://doi.org/10.1002/cphc.202300553

Göçügenci, A. & Keskin, S. (2023). The effect of graphene nanoplatelet addition on the mechanical, durability and self-healing properties of engineered cementitious composites. *MATEC Web of Conferences,* 378, 02024. https://doi.org/10.1051/matecconf/202337802024

Gopal, J., Muthu, M., & Sivanesan, I. (2023). A comprehensive compilation of graphene/fullerene polymer nanocomposites for electrochemical energy storage. *Polymers,* 15(3), 701. https://doi.org/10.3390/polym15030701

Gou, Y., Zhang, J., Liu, X., Zhou, Z., Zhang, M., Song, L., et al. (2024). A highly efficient additive for direct reactivation of waste $LiFePO_4$ with practical electrochemical performance. *Energy & Fuels,* 38(7), 6518–6527. https://doi.org/10.1021/acs.energyfuels.4c00329

Govindarajan, P., Antony, J., Palanisamy, S., Ayrılmış, N., Khan, T., Junaedi, H., et al. (2024). Advances in manufacturing of carbon-based molecular nanomaterials based on rice husk/hull waste. *Bioresources,* 19(4). https://doi.org/10.15376/biores.19.4.govindarajan

Goyal, M., Singh, K., & Bhatnagar, N. (2024). Applications of nanomaterials for enhanced performance, and sustainability in energy storage devices: a review. *ChemistrySelect,* 9(22). https://doi.org/10.1002/slct.202400543

Grandjean, M., Pichardo, M., Biecher, Y., Haon, C., & Chenevier, P. (2023). Matching silicon-based anodes with sulfide-based solid-state electrolytes for Li-ion batteries. *Journal of Power Sources,* 580, 233386. https://doi.org/10.1016/j.jpowsour.2023.233386

Guan, D., Wang, X., Song, L., Miao, C., Li, J., Yuan, X., et al. (2023). Polyoxometalate Li3PW12O40 and Li3PMo12O40 electrolytes for high-energy all-solid-state lithium batteries. *Angewandte Chemie,* 63(5). https://doi.org/10.1002/anie.202317949

Guan, Y., Li, W., Zhu, Y., Wang, X., Zhang, Y., & Chen, H. (2024). Compressed air energy storage system: effect of variable relative stator installation angle on nozzle governing turbine performance. *Journal of Physics: Conference Series,* 2707(1), 012125. https://doi.org/10.1088/1742-6596/2707/1/012125

Gudavalli, G., Nepal, M., Young, J., Smeu, M., & Dhakal, T. (2023). Effect of a Ti layer on the growth of binder-free carbon nanotubes on Cu foil and their performance as lithium-ion battery anodes. *Energy Advances,* 2(8), 1182–1189. https://doi.org/10.1039/d3ya00212h

Guo, D., Mu, L., Lin, F., & Liu, G. (2023). Mesoporous polyimide thin films as dendrite-suppressing separators for lithium–metal batteries. *ACS Nano,* 18(1), 155–163. https://doi.org/10.1021/acsnano.3c04159

Guo, J., Pan, L., Sun, J., Wei, D., Dai, Q., Xu, J., et al. (2023). Metal-free fabrication of nitrogen-doped vertical graphene on graphite felt electrodes with enhanced reaction kinetics and mass transport for high-performance redox flow batteries. *Advanced Energy Materials,* 14(1). https://doi.org/10.1002/aenm.202302521

Guo, Y., Zhao, J., Wang, C., Gu, C., & Wang, Y. (2023). High-temperature oxidation behavior of medium-manganese austenitic steel. *Steel Research International,* 94(12). https://doi.org/10.1002/srin.202300216

Guo, Z., Xu, Z., Xie, F., Jiang, J., Zheng, K., Alabidun, S., et al. (2023). Investigating the superior performance of hard carbon anodes in sodium-ion compared with lithium- and potassium-ion batteries. *Advanced Materials,* 35(42). https://doi.org/10.1002/adma.202304091

Hammoodi, O., Mohammed, I., Majeed, A., Alheety, M., Mohammed, L., & Yahya, M. (2024). Enhanced hydrogen storage capacity of graphene oxide through doping with copper ferrite nanoparticles. *Energy Storage*, 6(1). https://doi.org/10.1002/est2.574

Han, M., Mu, Y., Wei, L., Zeng, L., & Zhao, T. (2023). Multilevel carbon architecture of subnanoscopic silicon for fast-charging high-energy-density lithium-ion batteries. *Carbon Energy*, 6(4). https://doi.org/10.1002/cey2.377

Han, X., Qiao, L., Peng, P., Fu, C., Zhang, S., Liu, K., et al. (2024). Nickel nanoparticles encapsulated in N-doped carbon nanotubes for microwave absorption. *ACS Applied Nano Materials*, 7(3), 2593–2604. https://doi.org/10.1021/acsanm.3c04372

Hana, N. & Munasir, M. (2023). Study of performance graphene oxide modification of $LiFePO_4$/C material for the cathode of Li-ion batteries. *Journal of Physics Conference Series*, 2623(1), 012014. https://doi.org/10.1088/1742-6596/2623/1/012014

Harini, U. & Ponraj, C. (2024). Review—graphene-based electrochemical sensors for detecting environmental pollutants. *Journal of the Electrochemical Society*, 171(2), 027521. https://doi.org/10.1149/1945-7111/ad26e3

Harizanova, S., Vulchev, V., & Stoyanova, R. (2023). Graphene-based composites for thermoelectric applications at room temperature. *Materials*, 16(23), 7262. https://doi.org/10.3390/ma16237262

Hasan, A. & Susan, M. (2024). Synergism in carbon nanotubes and carbon-dots: counter electrode of a high-performance dye-sensitized solar cell. *RSC Advances*, 14(11), 7616–7630. https://doi.org/10.1039/d4ra00601a

Hasan, M., Hossain, S., Mofijur, M., Kabir, Z., Badruddin, I., Khan, T., et al. (2023). Harnessing solar power: a review of photovoltaic innovations, solar thermal systems, and the dawn of energy storage solutions. *Energies*, 16(18), 6456. https://doi.org/10.3390/en16186456

He, M., Lu, J., Shi, C., Qian, X., Yin, L., Gong, K., et al. (2023). Multi-stimuli-responsive aerogels composed of ag nanoparticle-coated graphene nanosheets for energy storage and conversion. *ACS Applied Nano Materials*, 6(18), 16503–16514. https://doi.org/10.1021/acsanm.3c02622

He, T., Zhao, W., Hu, J., Deng, C., Yan, D., & Huang, S. (2023). Unveiling the double-edged behavior of controlled selenium substitution in cobalt sulfide for balanced Na-storage capacity and rate capability. *Advanced Functional Materials*, 34(8). https://doi.org/10.1002/adfm.202310256

He, W., Shi, J., Shen, J., Che, J., & Zhang, Y. (2024). Vertical-channeled graphene oxide/cellulose composite flexible electrodes. *ACS Applied Nano Materials*, 7(22), 25569–25579. https://doi.org/10.1021/acsanm.4c04687

He, Y., Shen, X., & Zhang, Y. (2024). Layered 2D materials in batteries. *ACS Applied Nano Materials*, 7(24), 27907–27939. https://doi.org/10.1021/acsanm.3c06242

Helal, H., Ahrouch, M., Rabehi, A., Zappa, D., & Comini, E. (2024). Nanostructured materials for enhanced performance of solid oxide fuel cells: a comprehensive review. *Crystals*, 14(4), 306. https://doi.org/10.3390/cryst14040306

Hossain, M., Hamanah, W., Alam, M., Shafiullah, M., & Abido, M. (2023). Fault ride through and intermittency improvement of renewable energy integrated MMC-HVDC system employing flywheel energy storage. *IEEE Access*, 11, 50528–50546. https://doi.org/10.1109/access.2023.3277968

Hossain, M., Khan, M., Halim, M., Elme, N., & Hussain, M. (2023). A review on stability challenges and probable solution of perovskite–silicon tandem solar cells. *Signal and Image Processing Letters*, 5(1), 62–71. https://doi.org/10.31763/simple.v5i1.58

Hu, H., Zhong, J., Jian, B., Zheng, C., Zeng, Y., Kou, C., et al. (2024). In-situ construction of anti-aggregation tellurium nanorods/reduced graphene oxide composite to enable fast sodium storage. *Nanomaterials*, 14(1), 118. https://doi.org/10.3390/nano14010118

Hu, K., Sang, X., Chen, J., Liu, Z., Zhang, J., & Hu, X. (2023). High-safety lithium-ion batteries with silicon-based anodes enabled by electrolyte design. *Chemistry - An Asian Journal*, 18(24). https://doi.org/10.1002/asia.202300820

Hu, Z., Wang, S., Lynch, J., & Jariwala, D. (2024). Tandem photovoltaics from 2D transition metal dichalcogenides on silicon. *ACS Photonics*, 11(11), 4616–4625. https://doi.org/10.1021/acsphotonics.4c00982

Huang, J., Guo, Z., Xiao, Q., Zeng, M., & Zhong, S. (2024). Effect of S-doped carbon nanotubes as a positive conductive agent in lithium-ion batteries. *Ionics*, 30(6), 3187–3194. https://doi.org/10.1007/s11581–024-05541-1

Huang, L., Gu, Z., He, W., Shi, K., Peng, L., Sheng, Z., et al. (2023). Solvothermal synthesis and pyrolysis toward heteroatom-doped carbon microspheres for zinc-ion hybrid capacitors. *Small*, 20(14). https://doi.org/10.1002/smll.202308788

Huang, S., Lin, W., & Rajagopal, V. (2023). A comparative analysis of hydrogen storage characteristics in AZ31 magnesium alloy with the addition of graphene and carbon nanotubes via ball milling process. *Journal of Research Updates in Polymer Science*, 12, 231–237. https://doi.org/10.6000/1929-5995.2023.12.22

Hussein, A & Yassir, Y. (2024). Graphene as a promising material in orthodontics: a review. *Journal of Orthodontic Science*, 13(1) https://doi.org/10.4103/jos.jos_3_24

Hussein, H. (2023). The state of the art of nanomaterials and its applications in energy saving. *Bulletin of the National Research Centre*, 47(1). https://doi.org/10.1186/s42269-023-00984-4

Hwang, C., Hwang, J., Kwak, M., Kim, H., Song, H., & Song, W. (2024). Highly flexible electrodes based on nano/micro-fiber for flexible lithium metal batteries. *Advanced Functional Materials*, 34(45). https://doi.org/10.1002/adfm.202404649

Hwang, S. & Hong, D. (2023). Performance evaluation of composite electrolyte with GQD for all-solid-state lithium batteries. *Computers Materials & Continua*, 74(1), 55–66. https://doi.org/10.32604/cmc.2023.028845

Iijima, S. (1991). Synthesis of carbon nanotubes. *Nature*, 354, 56–58. https://doi.org/10.1038/354056a0

Ikinci, M., Zehni, O., & Bissett, M. (2024). Structural supercapacitors based on graphene nanoplatelet-coated carbon fiber electrodes. *ACS Applied Nano Materials*, 7(21), 24875–24885. https://doi.org/10.1021/acsanm.4c04801

Islam, T., Hasan, M., Sarker, S., & Ahammad, A. (2023). Intrinsic properties of GO/rGO bilayer electrodes dictate their inter-/intralayer intractability to modulate their capacitance performance. *ACS Omega*, 8(15), 14013–14024. https://doi.org/10.1021/acsomega.3c00573

Iqbal, M., Liu, Z., Hegazy, H., & Alahmari, A. (2024). Exploring the potential of solution-processable $AgBiS_2$ quantum dots as an efficient photoactive material for eco-friendly solar cell devices. *Energy & Fuels*, 38(24), 23746–23753. https://doi.org/10.1021/acs.energyfuels.4c04796

Jain, P., Al, R., Kumar, S., Sharma, A., Jain, A., Bora, B., et al. (2024). Recent advances in graphene-enabled materials for photovoltaic applications: a comprehensive review. *ACS Omega*. https://doi.org/10.1021/acsomega.3c07994

Jaiswal, N., Singh, V., Kumari, D., & Pandey, S. (2023). Performance analysis of triple-cation mixed-halide bifacial perovskite solar cell with 17% rear and 25% front efficiency. *Energy Technology*, 12(2). https://doi.org/10.1002/ente.202300639

Jamaluddin, A., Nursanti, A., Nur'aini, A., Putri, R., & Arshad, M. (2023). Graphene as an active material for supercapacitors: a machine learning approach. *Indonesian Journal of Applied Physics*, 13(2), 305. https://doi.org/10.13057/ijap.v13i2.76678

Jasiński, M. (2024). Advances in plasma and laser engineering. *Materials*, 17(8), 1768. https://doi.org/10.3390/ma17081768

Jauja-Ccana, V., Riveros, L., Cordova-Huaman, A., Huayta, G., Manfredy, L., Naupa, A., et al. (2023). Review—rational design of nitrogen-doped graphene as anode material for lithium-ion batteries. *Journal of the Electrochemical Society*, 170(4), 040525. https://doi.org/10.1149/1945-7111/accab0

Jawahar, D. & Pillai, A. (2023). Generation of green carbon nanomaterials from biomass pyrolysis for the production of biofuel and bioenergy. *Migration Letters*, 20(S13), 133–140. https://doi.org/10.59670/ml.v20is13.6276

Jeong, S., Ohto, T., Nishiuchi, T., Nagata, Y., Fujita, J., & Ito, Y. (2023). Suppression of methanol and formate crossover through sulfanilic-functionalized holey graphene as proton exchange membranes. *Advanced Science*, 10(31). https://doi.org/10.1002/advs.202304082

Ji, H. (2023). Performance and modification of novel nanomaterials for the anode of lithium-ion batteries. *Applied and Computational Engineering,* 26(1), 43–49. https://doi.org/10.54254/2755-2721/26/20230792

Jia, Q. (2023). Synthesis and modification of carbon nanotubes. *Applied and Computational Engineering,* 26(1), 115–119. https://doi.org/10.54254/2755-2721/26/20230810

Jiang, J., Cui, X., Huang, Y., Yan, D., Wang, B., Yang, Z., et al. (2024). Advances and prospects in integrated nano-oncology. *Nano Biomedicine and Engineering,* 16(2), 152–187. https://doi.org/10.26599/nbe.2024.9290060

Jiao, K., Liu, S., Ma, Y., Chen, S., Wei, L., & Yan, J. (2023). Long-term cycling quasi-solid-state lithium batteries enabled by 3D nanofibrous TiO_2–x@ Li anodes and in-situ polymerized gel-electrolytes. *Chemical Engineering Journal,* 464, 142627. https://doi.org/10.1016/j.cej.2023.142627

Jiao, L., Li, Z., Zhu, Y., Zheng, W., Liang, Y., Wang, X., et al. (2023). Enhanced electrical conductivity and lithium-ion diffusion rate of $LiFePO_4$ by Fe site and P site doping. *AIP Advances,* 13(7). https://doi.org/10.1063/5.0159212

Jiao, M., Du, J., Xu, H., Li, J., Fu, M., Yang, W., et al. (2024). Balancing charging efficiency and thermal safety: a comparative analysis of multistage constant current charging protocols. *Industrial & Engineering Chemistry Research,* 63(22), 10054-10066. https://doi.org/10.1021/acs.iecr.4c00971

Jitapunkul, K., Deshsorn, K., Payakkachon, K., Chaisrithong, T., Lawtrakul, L., & Iamprasertkun, P. (2023). Insights into heteroatom-doped graphene supercapacitor data through manual data separation and statistical analysis. *The Journal of Physical Chemistry C,* 127(37), 18316–18326. https://doi.org/10.1021/acs.jpcc.3c04956

Jo, S., Ramkumar, R., & Lee, J. (2023). Recent advances in laser-induced graphene-based materials for energy storage and conversion. *ChemSusChem,* 17(5). https://doi.org/10.1002/cssc.202301146

Journet, J., Maser, W.K., Bernier, P., Loiseau, A., de la Chapelle, M.L., Lefrant, S., Deniard, P., Lee, R., & Fisher, J.E. (1997). Large-scale production of single-walled carbon nanotubes by the electric-arc technique. *Nature,* 388, 756–758.

Julien, C. & Mauger, A. (2024). Fabrication of $Li_4Ti_5O_{12}$ (LTO) as anode material for Li-ion batteries. *Micromachines,* 15(3), 310. https://doi.org/10.3390/mi15030310

Kalita, D., Sahu, P., & Manju, U. (2024). Anti-perovskites for photovoltaics: materials development and challenges. *Journal of Physics D: Applied Physics,* 57(34), 343002. https://doi.org/10.1088/1361-6463/ad4daf

Kang, C., Lin, L., Nguyen, T., Chen, C., Chang, J., Lin, T., et al. (2024). High-power battery electrodes fabricated by acupuncture-inspired microneedle processing. *ACS Applied Materials & Interfaces.* https://doi.org/10.1021/acsami.4c11834

Kang, T., Lee, J., Lee, J., Park, J., Shin, J., Ju, J., et al. (2023). An ion-channel-restructured zwitterionic covalent organic framework solid electrolyte for all-solid-state lithium-metal batteries. *Advanced Materials,* 35(30). https://doi.org/10.1002/adma.202301308

Kashyap, B., Singh, V., Solanki, M., Kumar, A., Ruokolainen, J., & Kesari, K. (2023). Smart nanomaterials in cancer theranostics: challenges and opportunities. *ACS Omega,* 8(16), 14290–14320. https://doi.org/10.1021/acsomega.2c07840

Kashyap, R. & Pillai, P. (2024). Plasmonic nanoparticles boost solar-to-electricity generation at ambient conditions. *Nano Letters,* 24(18), 5585–5592. https://doi.org/10.1021/acs.nanolett.4c00925

Kausar, A. (2024). Energy systems and green sourced nanomaterials—a today's outlook. *Mater. Tech. Rep.,* 2(2), 1879. https://doi.org/10.59400/mtr1879

Kausar, A. & Ahmad, I. (2023). High-tech graphene oxide reinforced conducting matrix nanocomposites—current status and progress. *Characterization and Application of Nanomaterials,* 6(1). https://doi.org/10.24294/can.v6i1.2637

Kausar, A., Ahmad, I., Zhao, T., Aldaghri, O., Ibnaouf, K., & Eisa, M. (2023). Graphene nanocomposites as innovative materials for energy storage and conversion—design and headways. *International Journal of Molecular Sciences,* 24(14), 11593. https://doi.org/10.3390/ijms241411593

Khakpour, E., Salehi, S., Naghib, S., Ghorbanzadeh, S., & Zhang, W. (2023). Graphene-based nanomaterials for stimuli-sensitive controlled delivery of therapeutic molecules. *Frontiers in Bioengineering and Biotechnology,* 11. https://doi.org/10.3389/fbioe.2023.1129768

Khan, M. (2024). Innovations in battery technology: enabling the revolution in electric vehicles and energy storage. *British Journal of Multidisciplinary and Advanced Studies,* 5(1), 23–41. https://doi.org/10.37745/bjmas.2022.0414

Khan, M., Tahir, D., Asim, M., Israr, M., Haider, A., & Xu, D. (2024). Revolutionizing cancer care: advances in carbon-based materials for diagnosis and treatment. *Cureus.* https://doi.org/10.7759/cureus.52511

Khodadadi, A., Zhang, P., Wang, J., Guo, Z., Yang, G., & Xu, H. (2024). Materials and energy applications of graphene and its composites. *Advanced Materials Science,* 17(2), 345–368.

Kim, J., Lee, S., Park, S., Ju, M., Kim, Y., Cho, E., et al. (2023). Highly efficient bifacial silicon/silicon tandem solar cells. *IEEE Access,* 11, 21326–21331. https://doi.org/10.1109/access.2023.3248795

Kim, J., Ling, J., Lai, Y., & Milner, P. (2024). Redox-active organic materials: from energy storage to redox catalysis. *ACS Materials Au,* 4(3), 258–273. https://doi.org/10.1021/acsmaterialsau.3c00096

Kishore, S., Perumal, S., Atchudan, R., Alagan, M., Wadaan, M., Baabbad, A., et al. (2023). Recent advanced synthesis strategies for the nanomaterial-modified proton exchange membrane in fuel cells. *Membranes,* 13(6), 590. https://doi.org/10.3390/membranes13060590

Kothandam, G., Singh, G., Guan, X., Lee, J., Ramadass, K., Joseph, S., et al. (2023). Recent advances in carbon-based electrodes for energy storage and conversion. *Advanced Science,* 10(18). https://doi.org/10.1002/advs.202301045

Kubarkov, A., Babkin, A., Drozhzhin, O., Stevenson, K., Antipov, E., & Sergeyev, V. (2023). Engendering high energy density $LiFePO_4$ electrodes with morphological and compositional tuning. *Nanomaterials,* 13(11), 1771. https://doi.org/10.3390/nano13111771

Kumar, N., Chaudhuri, A., Arya, V., Bakli, C., & Bera, C. (2023). Significantly reduced thermal conductivity and enhanced thermoelectric performance of twisted bilayer graphene. *Journal of Applied Physics,* 134(4). https://doi.org/10.1063/5.0153052

Kumar, S., Ding, W., Hoffmann, R., Sieuw, L., Heinz, M., Weber, N., et al. (2023). $AlCl_3$-NaCl-$ZnCl_2$ secondary electrolyte in next-generation zebra (Na-$ZnCl_2$) battery. *Batteries*, 9(8), 401. https://doi.org/10.3390/batteries9080401

Laconi, N., Licheri, F., Cambuli, F., Cocco, D., & Petrollese, M. (2024). Optimized design procedure for small-scale pumped hydro energy storage systems based on the use of pumps as turbines. *Journal of Physics Conference Series,* 2893(1), 012011. https://doi.org/10.1088/1742-6596/2893/1/012011

Lahoz, R., Naghilou, A., Kautek, W., & Bomati-Miguel, O. (2020). Study of the physicochemical surface alterations and incubation phenomena induced on iron targets by nanosecond pulsed laser ablation in liquids: effect on productivity and characteristics of the synthesized nanoscale zero-valent iron (nZVI) particles. *Applied Surface Science*, 511, 145438. https://doi.org/10.1016/j.apsusc.2020.145438

Lai, W., Xu, X., Xi, K., Lei, Y., Cheng, X., Shi, X., et al. (2024). Ultralong cycling and interfacial regulation of bilayer heterogeneous composite solid-state electrolytes in lithium metal batteries. *ACS Applied Materials & Interfaces,* 16(26), 33578–33589. https://doi.org/10.1021/acsami.4c06026

Lama, B., Smirnova, A., & Paudel, T. (2023). Enhanced li-ion diffusivity of $LiFePO_4$ by Ru doping: ab initio and machine learning force field results. *ACS Applied Energy Materials*, 6(20), 10424–10431. https://doi.org/10.1021/acsaem.3c01429

Lan, J. (2023). Performance of LIBs with nanostructured silicon as anode materials. *Highlights in Science Engineering and Technology*, 43, 569–575. https://doi.org/10.54097/hset.v43i.7478

Lavi, A., Ohayon-Lavi, A., Leibovitch, Y., Hayun, S., Ruse, E., & Regev, O. (2023). Thermally conductive molten salt for thermal energy storage: synergistic effect of a hybrid graphite-graphene nanoplatelet filler. *Global Challenges*, 7(9). https://doi.org/10.1002/gch2.202300053

Lavorato, C. & Fontananova, E. (2023). An overview on exploitation of graphene-based membranes: from water treatment to medical industry, including recent fighting against covid-19. *Microorganisms*, 11(2), 310. https://doi.org/10.3390/microorganisms11020310

Lee, H., Choi, W., Park, H., & Kim, D. (2023). Realization of high loading density lithium polymer batteries by optimizing lithium-ion transport and electronic conductivity. *ACS Applied Materials & Interfaces*, 15(12), 15298–15310. https://doi.org/10.1021/acsami.2c20455

Lee, S., Koo, H., Kang, H., Oh, K., & Nam, K. (2023). Advances in polymer binder materials for lithium-ion battery electrodes and separators. *Polymers*, 15(23), 4477. https://doi.org/10.3390/polym15234477

Liao, Y., Lin, W., Zhang, Y., Yang, J., Li, Z., Ren, Y., et al. (2024). A weakly solvating ether electrolyte enables fast-charging and wide-temperature lithium-ion pouch cells. *ACS Nano*, 18(31), 20762–20771. https://doi.org/10.1021/acsnano.4c06997

Liang, C., Jiang, Q., Liu, Z., Yang, J., Zhang, J., Zhang, S., et al. (2023). Effect of sublethal concentrations of metal nanomaterials on cell energy metabolism. *Toxics*, 11(5), 453. https://doi.org/10.3390/toxics11050453

Liao, H. (2024). Research progress on the adsorption of pollutants in water by oxidized graphene. *Science and Technology of Engineering, Chemistry and Environmental Protection*, 1(8). https://doi.org/10.61173/f19bfp84

Liao, Y., Lin, W., Zhang, Y., Yang, J., Li, Z., Ren, Y., et al. (2024). A weakly solvating ether electrolyte enables fast-charging and wide-temperature lithium-ion pouch cells. *ACS Nano*, 18(31), 20762–20771. https://doi.org/10.1021/acsnano.4c06997

Licheri, F., Petrollese, M., Cocco, D., & Cambuli, F. (2023). Pumps as turbines for pumped hydro energy storage systems - a small-size case study. *Journal of Physics Conference Series*, 2648(1), 012046. https://doi.org/10.1088/1742-6596/2648/1/012046

Lin, G., Kang, Y., Li, G., Wu, J., Chen, Y., Liu, C., et al. (2024). Regenerating spent lithium-ion battery cathodes as dual-functional biocatalysts for highly-efficient oxygen and hydrogen evolution using *Aspergillus niger* soluble extracellular polymers. *ACS Sustainable Chemistry & Engineering*, 12(19), 7466–7477. https://doi.org/10.1021/acssuschemeng.4c00906

Lin, H., Chuang, E., & Lin, S. (2024). Advancing lithium battery performance through porous conductive polyaniline-modified graphene composites additive. *Nanomaterials*, 14(6), 509. https://doi.org/10.3390/nano14060509

Lin, H., Yang, J., Feng, P., Liu, G., Liu, Y., Cui, J., et al. (2023). Structure regulation and application of bagasse-based porous carbon material based on H_2O_2-assisted hydrothermal treatment. *Energy & Fuels*, 37(15), 11342–11354. https://doi.org/10.1021/acs.energyfuels.3c01458

Lin, J., Mo, Y., Li, Z., Yang, K., & Yu, J. (2023). Simultaneously encapsulating Co/Co_2P nanoparticles inside and growing vertical graphene nanosheets outside for N-doped carbon nanotubes as efficient sulfur carriers for high-performance lithium sulfur batteries. *Energy Technology*, 11(9). https://doi.org/10.1002/ente.202300283

Lin, L., Yu, L., Guo, C., Xiong, X., Fan, F., Yang, W., et al. (2023). Green fabrication of agglomeration-reductive and electrochemical-active reduced graphene oxide/polymerized proanthocyanidins electrode for high-performance supercapacitor. *ACS Applied Engineering Materials*, 1(1), 369–379. https://doi.org/10.1021/acsaenm.2c00090

Lin, W., Suresh, G., Gupta, V., Tomida, S., Oyama, K., Yaakob, Y., et al. (2024). In situ transmission electron microscopy observation of electrochemical process between Li–C nanocomposites and multilayer graphene. *Advanced Materials Technologies*, 9(5). https://doi.org/10.1002/admt.202301564

Lin, Z. & Wang, Z. (2023). Application of solid polymer electrolytes for solid-state sodium batteries. *Matec Web of Conferences*, 386, 03019. https://doi.org/10.1051/matecconf/202338603019

Liu, F., Hu, Y., Qu, Z., Ma, X., Li, Z., Zhu, R., et al. (2023). Rapid production of kilogram-scale graphene nanoribbons with tunable interlayer spacing for an array of renewable energy. *Proceedings of the National Academy of Sciences*, 120(26). https://doi.org/10.1073/pnas.2303262120

Liu, G., Li, X., Li, C., Zheng, Q., Wang, Y., Xiao, R., et al. (2024). Efficient fabrication of disordered graphene with improved ion accessibility, ion conductivity, and density for high-performance compact capacitive energy storage. *Advanced Science*, 11(38). https://doi.org/10.1002/advs.202405155

Liu, G., Liu, X., Ma, X., Tang, X., Zhang, X., Dong, J., et al. (2023). High-performance dual-ion battery based on silicon–graphene composite anode and expanded graphite cathode. *Molecules*, 28(11), 4280. https://doi.org/10.3390/molecules28114280

Liu, H. (2024). The research progress on photocatalytic performance of graphene-based nanocomposites. *Applied and Computational Engineering*, 61(1), 1–5. https://doi.org/10.54254/2755-2721/61/20240905

Liu, H., Liu, X., Wang, Z., Zhu, L., & Zhang, X. (2023). Evolution process of the interfacial chemical reaction in Ni-rich layered cathodes for all-solid-state batteries. *ACS Applied Materials & Interfaces*, 16(1), 943–956. https://doi.org/10.1021/acsami.3c16689

Liu, H., Li, N., Zhang, S., Wang, J., Du, Y., & Zhang, W. (2023). Design of gradient Ti reconstituted Fe_2O_3 anodes with enhanced lithium affinity modulated electronic structures: first-principles calculations and experiment verification. *ACS Applied Materials & Interfaces*, 15(19), 23160–23169. https://doi.org/10.1021/acsami.3c02028

Liu, J. (2023a). Progress in the application of graphene nanomaterials in the cement-based composites. *Juniper Online Journal Material Science*, 7(5). https://doi.org/10.19080/jojms.2023.07.555723

Liu, J. (2023b). Application of nanomaterials in the negative electrode of lithium-ion batteries. *Applied and Computational Engineering*, 24(1), 246–250. https://doi.org/10.54254/2755-2721/24/20230716

Liu, L., Zhang, Y., Xue, N., Wang, Y., Wang, R., Wang, L., et al. (2024). Mitigation of volumetric expansion in silicon anodes via engineered porosity: electrochemical performances and stress distribution implication. *Energy Technology*, 12(9). https://doi.org/10.1002/ente.202400664

Liu, M., Zhang, B., Wang, J., Liu, H., Wang, J., Liu, C., et al. (2023). Optimal configuration of wind-PV and energy storage in large clean energy bases. *Sustainability*, 15(17), 12895. https://doi.org/10.3390/su151712895

Liu, M., Wang, L., & Yu, G. (2024). Recent research progress of porous graphene and applications in molecular sieve, sensor, and supercapacitor. *Small*, 20(42). https://doi.org/10.1002/smll.202401767

Liu, Q., Ge, K., Wu, X., Zhu, Z., Zhu, Y., & Bi, H. (2023). Preparation of carbon dots@r-GO nanocomposite with an enhanced pseudo-capacitance. *Molecules*, 28(2), 541. https://doi.org/10.3390/molecules28020541

Liu, S., Xiao, D., Wang, B., Wang, L., Wu, T., Wang, Y., et al. (2023). High-voltage spinel and Li_2MnO_3 composite structure construction in $LiMn_{0.8}Ni_{0.2}O_2$ for manganese-based lithium-ion battery cathode materials. *Advanced Energy Materials*, 13(19). https://doi.org/10.1002/aenm.202300217

Liu, S., Fan, B., Shi, Z., Wan, R., Sheng, X., Li, X., et al. (2024). High-safety lithium-ion battery separator with adjustable temperature function inspired by the sugar gourd structure. *ACS Applied Materials & Interfaces*, 16(23), 30284–30295. https://doi.org/10.1021/acsami.4c04937

Liu, T., Yang, Y., Cao, S., Xiang, R., Zhang, L., & Yu, J. (2023). Pore perforation of graphene coupled with in situ growth of Co_3Se_4 for high-performance Na-ion battery. *Advanced Materials*, 35(13). https://doi.org/10.1002/adma.202207752

Liu, X. & Huang, Z. (2024). Rapid treatment of LLZTO solid electrolyte surface impurities by lactic acid solution. *Journal of Physics Conference Series*, 2840(1), 012018. https://doi.org/10.1088/1742-6596/2840/1/012018

Liu, X., Liu, G., Fu, T., Ding, K., Guo, J., Wang, Z., et al. (2024). Structural design and energy and environmental applications of hydrogen-bonded organic frameworks: a systematic review. *Advanced Science*, 11(22). https://doi.org/10.1002/advs.202400101

Liu, X. (2024). Graphene's potential in electronic devices: characteristics, applications, challenges, and solutions. *Highlights in Science Engineering and Technology*, 81, 478–483. https://doi.org/10.54097/2pfayn35

Liu, Y., Tan, X., Liang, J., Han, H., Xiang, P., & Yan, W. (2023). Machine learning for perovskite solar cells and component materials: key technologies and prospects. *Advanced Functional Materials*, 33(17). https://doi.org/10.1002/adfm.202214271

Liu, Y., Cui, C., Deng, L., Wang, Z., Dong, L., Dong, H., et al. (2023). Magnesium anchoring strategy for stabilizing graphene-hosted lithium metal battery. *Small Structures*, 5(2). https://doi.org/10.1002/sstr.202300345

Liu, Y., Wang, C., Yoon, S., Han, S., Lewis, J., Prakash, D., et al. (2023). Aluminum foil negative electrodes with multiphase microstructure for all-solid-state li-ion batteries. *Nature Communications*, 14(1). https://doi.org/10.1038/s41467-023-39685-x

Liu, Z., Shu, K., Yang, Y., Yu, H., Huang, Y., Gao, W., et al. (2023). Integrating graphene enables improved and gate-tunable photovoltaic effect in Van der Waals heterojunction. *Advanced Optical Materials*, 11(6). https://doi.org/10.1002/adom.202202646

Liu, Z., Zhang, S., Zhou, Q., Zhang, Y., Lv, D., Shen, Y., et al. (2023). Insights into quasi solid-state polymer electrolyte: the influence of succinonitrile on polyvinylene carbonate electrolyte in view of electrochemical applications. *Battery Energy*, 2(3). https://doi.org/10.1002/bte2.20220049

Liu, Z., Liu, R., Xu, S., Tian, J., Li, J., Li, H., et al. (2024). Achieving a deeply desodiated stabilized cathode material by the high entropy strategy for sodium-ion batteries. *Angewandte Chemie*, 63(29). https://doi.org/10.1002/anie.202405620

Luan, D. (2023). Applications of graphene in different fields. *Matec Web of Conferences*, 386, 03015. https://doi.org/10.1051/matecconf/202338603015

Luo, D., Shi, T., Li, Q., Xu, Q., Strømme, M., Zhang, Q., et al. (2023). Green, general and low-cost synthesis of porous organic polymers in sub-kilogram scale for catalysis and co2 capture. *Angewandte Chemie*, 135(27). https://doi.org/10.1002/ange.202305225

Lye, Y., Chan, K., & Ng, Z. (2023). A review on the progress, challenges, and performances of tin-based perovskite solar cells. *Nanomaterials*, 13(3), 585. https://doi.org/10.3390/nano13030585

Ma, L. (2024). Improving energy density in lithium-ion batteries using nanomaterials. *Highlights in Science Engineering and Technology*, 116, 142–148. https://doi.org/10.54097/nfkgp156

Ma, H., Wang, F., Shen, M., Tong, Y., Wang, H., & Hu, H. (2023). Advances of $LiCoO_2$ in cathode of aqueous lithium-ion batteries. *Small Methods*, 8(6). https://doi.org/10.1002/smtd.202300820

Machado, A., Schmitt, P., Maraschin, T., Osório, D., Basso, N., & Berlese, D. (2024). Adsorption capacity of pollutants from water by graphene and graphene-based materials: a bibliographic review. *Contribuciones a Las Ciencias Sociales*, 17(2), e4707. https://doi.org/10.55905/revconv.17n.2-285

Mahajani, V., Bhimani, K., Lakhnot, A., Panchal, R., Anjan, A., Manoj, R., et al. (2024). Facile magnetite coating for stable high-voltage cycling of nickel-rich cathodes in conventional liquid and all-solid-state lithium-ion batteries. *Small*, 20(33). https://doi.org/10.1002/smll.202402126

Mahalingam, A., Hemalatha, P., & Deivanayagam, P. (2024). Fabrication of zinc metal organic framework-embedded electrolyte membranes comprising sulfonated poly(ether ether sulfone) and polyimide for fuel cells. *Industrial & Engineering Chemistry Research*, 63(44), 19007–19016. https://doi.org/10.1021/acs.iecr.4c03412

Mahmood, R. (2025). Enhanced electrochemical performance of γ-Fe_2O_3/Cr_2O_3/graphene oxide/sodium alginate hybrid nanocomposite aerogels for high-efficiency ultracapacitors. *Journal of Nanotechnology*, 2025(1). https://doi.org/10.1155/jnt/8176379

Mahmood, S., Hussien, F., & Atiyah, A. (2023). Carbon nanotubes: synthesis via flame fragment deposition (FFD) method from liquefied petroleum gas. *Baghdad Science Journal*, 20(2), 0287. https://doi.org/10.21123/bsj.2022.6966

Mahmud, M. (2023). A concise review of nanoparticles utilized energy storage and conservation. *Journal of Nanomaterials*, 2023, 1–14. https://doi.org/10.1155/2023/5432099

Mai, S., Zhang, W., Mu, X., & Cao, J. (2024). Structural decoration of porphyrin/phthalocyanine photovoltaic materials. *ChemSusChem*, 17(15). https://doi.org/10.1002/cssc.202400217

Majó, M., Svobodova-Sedlackova, A., Fernández, A., Calderón, A., & Barreneche, C. (2024). Thermal cycling test of solar salt in contact with sustainable solid particles for concentrating solar power (CSP) plants. *Energies*, 17(10), 2349. https://doi.org/10.3390/en17102349

Malik, S., Muhammad, K., & Waheed, Y. (2023). Emerging applications of nanotechnology in healthcare and medicine. *Molecules*, 28(18), 6624. https://doi.org/10.3390/molecules28186624

Manzoor, S., Ali, S., Mansha, M., Sadaqat, M., Ashiq, M., Tahir, M., et al. (2024). Exploring nanomaterials for hydrogen storage: advances, challenges, and perspectives. *Chemistry - An Asian Journal*, 19(16). https://doi.org/10.1002/asia.202400365

Marfoua, B. & Hong, J. (2023). Graphene induced high thermoelectric performance in ZnO/graphene heterostructure. *Advanced Materials Interfaces*, 10(7). https://doi.org/10.1002/admi.202202387

Martín-Folgar, R., Esteban-Arranz, A., Negri, V., & Morales, M. (2023). Graphene oxides (GOs) with different lateral dimensions and thicknesses affect the molecular response in *Chironomus riparius*. *Nanomaterials*, 13(6), 967. https://doi.org/10.3390/nano13060967

Masenga, B., Byiringiro, J., Kagiri, C., Tafara, E., Ngoma, D., & Aristid, G. (2023). Design and development of wind-solar hybrid power system with compressed air energy storage for voltage and frequency regulations. *Journal of Power and Energy Engineering*, 11(02), 1–24. https://doi.org/10.4236/jpee.2023.112001

Massimi, S., Witteman, L., Gauderman, K., & Rippy, K. (2023). Assessment of galvanic corrosion in molten salts. *SolarPACES Conf Proc*, 1. https://doi.org/10.52825/solarpaces.v1i.654

Matos, C., Silva, P., & Carneiro, J. (2023). Economic assessment for compressed air energy storage business model alternatives. *Applied Energy*, 329, 120273. https://doi.org/10.1016/j.apenergy.2022.120273

Maulana, M., Lee, H., Gyan-Barimah, C., Sung, J., & Yu, J. (2024). Hollow PtCo alloy nanostructures for efficient oxygen reduction electrocatalysis in polymer electrolyte membrane fuel cells. *Journal of Materials Chemistry A*, 12(41), 27979–27986. https://doi.org/10.1039/d4ta04470c

Menassol, G., Dubois, L., Nadolska, M., Vadgama, P., Martin, D., & Zebda, A. (2023). A biocompatible iron doped graphene based cathode for an implantable glucose biofuel cell. *Electrochimica Acta*, 439, 141627. https://doi.org/10.1016/j.electacta.2022.141627

Meng, J. (2024). Nanomaterial modified perovskite solar cells: black phosphorus, 2D perovskite materials, and quantum dots. *Highlights in Science Engineering and Technology*, 111, 126–129. https://doi.org/10.54097/f4p7v973

Meng, L., Xu, Q., Zhang, J., & Wang, X. (2024). Colloidal quantum dot materials for next-generation near-infrared optoelectronics. *Chemical Communications*, 60(9), 1072–1088. https://doi.org/10.1039/d3cc04315k

Meng, X., Peng, Q., Wen, J., Song, K., Peng, L., Wu, T., et al. (2023). Sulfonated poly (ether ether ketone) membranes for vanadium redox flow battery enabled by the incorporation of ionic liquid-covalent organic framework complex. *Journal of Applied Polymer Science*, 140(18). https://doi.org/10.1002/app.53802

Mo, S., Xiao, B., Mo, B., Chen, J., Jia, L., Wang, Z., et al. (2023). Improving the thermal and photothermal performances of MXene-doped microencapsulated molten salts for medium-temperature solar thermal energy storage. *Energy & Fuels*, 37(10), 7490–7500. https://doi.org/10.1021/acs.energyfuels.3c00508

Modi, S., Yadav, V., Ali, D., Choudhary, N., Alarifi, S., Sahoo, D., et al. (2023). Photocatalytic degradation of methylene blue from aqueous solutions by using nano-ZnO/kaolin-clay-based nanocomposite. *Water*, 15(22), 3915. https://doi.org/10.3390/w15223915

Moghaddam, N., Akbari, N., Nandy, S., Chae, K., & Najafpour, M. (2023). Effect of different metal ions between nanolayers of manganese oxide for water oxidation reaction under acidic conditions. *The Journal of Physical Chemistry C*, 127(20), 9529–9540. https://doi.org/10.1021/acs.jpcc.3c01760

Mombeshora, E., Muchuweni, E., & Hashemi, H. (2024). Applications of graphene derivatives in all-solid-state supercapacitors. *ChemistrySelect*, 9(44). https://doi.org/10.1002/slct.202404345

Mose, P. (2024). The applications of nanotechnology in renewable energy. *Journal of Physical Sciences*, 5(1), 1–12. https://doi.org/10.47941/jps.1626

Mostafa, O., Zidan, N., Abbas, W., Issa, H., Gamal, N., & Fedawy, M. (2023). Design and performance optimization of lead-free perovskite solar cells with enhanced efficiency. *Mathematical Modelling and Engineering Problems*, 10(04), 1307–1316. https://doi.org/10.18280/mmep.100424

Moradi, R., Khalili, N.P., Khanmamadova, E., Abaszade, R. (2024). "Functionalized carbon nanostructures for flexible electronics," in *Handbook of Functionalized Carbon Nanostructures*. Cham: Springer, 2581–2614. https://doi.org/10.1007/978-3-031-14955-9_71-1

Mousa, A., Chuang, C., Kuo, S., & Mohamed, M. (2023). Strategic design and synthesis of ferrocene linked porous organic frameworks toward tunable CO_2 capture and energy storage. *International Journal of Molecular Sciences*, 24(15), 12371. https://doi.org/10.3390/ijms241512371

Mousavi, S., Hashemi, S., Kalashgrani, M., Gholami, A., Binazadeh, M., Chiang, W., et al. (2023). Recent advances in energy storage with graphene oxide for supercapacitor technology. *Sustainable Energy & Fuels*, 7(21), 5176–5197. https://doi.org/10.1039/d3se00867c

Mu, C., Qi, Y., Liu, W., & Bai, Y. (2023). Stabilizing commercial $LiMn_2O_4$ cathode by constructing protective saccharin coating. *ACS Applied Electronic Materials*, 5(3), 1793–1803. https://doi.org/10.1021/acsaelm.3c00009

Muduli, R., Nishad, N., Dashbabu, D., Kumar, E., & Kale, P. (2024). Synergistic integration of nickel, porous silicon, and thermally reduced graphene oxide for solid-state hydrogen energy storage. *Energy Storage*, 6(5). https://doi.org/10.1002/est2.70008

Mulla, R., White, A., Dunnill, C., & Barron, A. (2023). The role of graphene in new thermoelectric materials. *Energy Advances*, 2(5), 606–614. https://doi.org/10.1039/d3ya00085k

Nagaraja, T., Rajavel, K., Raihan, K., Lacroix, B., & Das, S. (2023). Microplotter-printed graphene-based electrochemical sensor for detecting phosphates. *ACS Applied Nano Materials*, 6(21), 20288–20297. https://doi.org/10.1021/acsanm.3c04228

Naghavi, N., Jalaly, M., Mohammadi, S., & Mousavi-Khoshdel, S. (2024). An investigation into the influence of graphene content on achieving a high-performance TiO2-graphene nanocomposite supercapacitor. *ChemistryOpen*, 13(11). https://doi.org/10.1002/open.202400128

Nandhakumar, M., Muthukrishnan, R., Kumar, T., Sureshkumar, P., & Nandini, S. (2023). Nano fertilizer in crop production: the changing scenario.

International Journal of Environment and Climate Change, 13(6), 158–170. https://doi.org/10.9734/ijecc/2023/v13i61811

Navarro, D., Alvira, D., & Manyà, J. (2023). Hard carbons from waste hemp via hydrothermal carbonization with mil chemical activation for sodium–ion batteries. *Jornadas De Jóvenes Investigadores Del I3A*, 11. https://doi.org/10.26754/jjii3a.20239012

Nazir, S., Ali, A., Aftab, A., Muqeet, H., Mirsaeidi, S., & Zhang, J. (2023). Techno-economic and environmental perspectives of solar cell technologies: a comprehensive review. *Energies*, 16(13), 4959. https://doi.org/10.3390/en16134959

Nazma, S., Sudha, T., Biradar, D., Krishnaraj, P., Chandrashekhar, S., & Ravikumar, H. (2024). Biosynthesized zinc nanoparticles and their plant growth-promoting effect on wheat (*Triticum aestivum* L.) cultivation. *Journal of Scientific Research and Reports*, 30(9), 148–158. https://doi.org/10.9734/jsrr/2024/v30i92339

Neira-Velázquez, M., Kú-Herrera, J., Narro-Céspedes, R., Flores-Villaseñor, S., Cortez-Garza, Y., Gaona, C., et al. (2024). Carbon nanostructures synthesis by catalyst-free atmospheric pressure plasma jet. *Journal of Physics D: Applied Physics*, 57(31), 315302. https://doi.org/10.1088/1361-6463/ad44a6

Nepal, M., Gudavalli, G., & Dhakal, T. (2025). MnO_2 nanoflowers electrodeposited on vertically aligned CNTs as binder-free electrodes for high-rate supercapacitors. *ACS Omega*, 10(4), 3439–3448. https://doi.org/10.1021/acsomega.4c07287

Nerath, G., Oliveira, D., Siqueira, J., & Caseli, L. (2024). Using carbon nanotubes to improve enzyme activity and electroactivity of fatty acid Langmuir–Blodgett film-incorporated galactose oxidase for sensing and energy storage applications. *ACS Applied Materials & Interfaces*, 17(9). https://doi.org/10.1021/acsami.3c18824

Nogales, P., Lee, S., Yang, S., Yang, I., Choi, S., Park, S., et al. (2024). Stabilizing the solid electrolyte interphase of SiOx negative electrodes: the role of fluoroethylene carbonate in enhancing electrochemical performance. *Batteries*, 10(11), 385. https://doi.org/10.3390/batteries10110385

Obayomi, K., Lau, S., Danquah, M., Zhang, J., Chiong, T., Takeo, M., et al. (2023). Novel concepts for graphene-based nanomaterials synthesis for phenol removal from palm oil mill effluent (POME). *Materials*, 16(12), 4379. https://doi.org/10.3390/ma16124379

Olcay, R., Palacios, E., Reyes, I., García, L., Ortega, P., Ordóñez, S., et al. (2024). Kinetics of thermal decomposition of carbon nanotubes decorated with magnetite nanoparticles. *C – Journal of Carbon Research*, 10(4), 96. https://doi.org/10.3390/c10040096

Ortiz, Y., Arévalo, P., Banegas, D., & Jurado, F. (2024). Recent advances in thermal management strategies for lithium-ion batteries: a comprehensive review. *Batteries,* 10(3), 83. https://doi.org/10.3390/batteries10030083

Osmani, P. (2024). Synthesis and characterization of advanced inorganic nanomaterials for energy storage devices. *Journal for Research in Applied Sciences and Biotechnology,* 3(1), 122–128. https://doi.org/10.55544/jrasb.3.1.20

Oza, P., Vasoya, N., & Vansdadiya, R. (2023). Empowering energy storage using graphene and its derivatives. *Current Journal of Applied Science and Technology,* 42(37), 24–33. https://doi.org/10.9734/cjast/2023/v42i374244

Pahlevaninezhad, M., Miller, E., Yang, L., Prophet, L., Singh, A., Storwick, T., et al. (2023). Exfoliated graphene composite membrane for the all-vanadium redox flow battery. *ACS Applied Energy Materials,* 6(12), 6505–6517. https://doi.org/10.1021/acsaem.3c00428

Palanichamy, P., Krishnasamy, R., Ulaganathan, M., Thiagamani, S., Ilyas, R., & Hassan, A. (2023). A practical green synthesis method of Ag NPs using rosy periwinkle plant leaves for solar panel coating. *Heliyon,* 9(12), e22893. https://doi.org/10.1016/j.heliyon.2023.e22893

Palanisamy, R., Muruganantham, B., Rajasekaran, S., Babu, B., Ramkumar, R., Marthanda, A., et al. (2024). A comprehensive review on different types of fuel cell and its applications. *Bulletin of Electrical Engineering and Informatics,* 13(2), 774–780. https://doi.org/10.11591/eei.v13i2.6348

Pan, C., He, J., Zhu, J., Li, S., Li, W., Yang, W., et al. (2024). Corrosion control by carbon-based nanomaterials: a review. *ACS Applied Nano Materials,* 7(3), 2515–2528. https://doi.org/10.1021/acsanm.3c05547

Pan, F. (2023). Application of nano-carbon materials in battery electrode. *Highlights in Science Engineering and Technology,* 43, 597–601. https://doi.org/10.54097/hset.v43i.7507

Pandey, B. & Goswami, Y. (2023). Optically important Cu_2ZnSnS_4 (CZTS) nanoparticles synthesis using a hydrothermal route with citric acid as a structure-directing agent. *Oriental Journal of Chemistry,* 39(4), 1033–1039. https://doi.org/10.13005/ojc/390429

Pandey, S., Karakoti, M., Bhardwaj, D., Tatrari, G., Sharma, R., Pandey, L., et al. (2023). Recent advances in carbon-based materials for high-performance perovskite solar cells: gaps, challenges and fulfillment. *Nanoscale Advances,* 5(6), 1492–1526. https://doi.org/10.1039/d3na00005b

Panneerselvam, V. & Sathian, S. (2024). Thermal transport in a defective pillared graphene network: insights from equilibrium molecular dynamics simulation. *Physical Chemistry Chemical Physics,* 26(14), 10650–10659. https://doi.org/10.1039/d4cp00147h

Panwar, V., Chauhan, P., Kumar, S., Tripathi, R., & Misra, A. (2023). Gate field induced extraordinary energy storage in MoS_2-graphene-based ultramicro-electrochemical capacitor. *ACS Energy Letters,* 8(3), 1510–1519. https://doi.org/10.1021/acsenergylett.2c02476

Park, W., Park, S., Cho, H., Oh, Y., Kang, I., & Ahn, J. (2024). In situ electrical contacts to graphene by laser scribing. *ACS Applied Nano Materials,* 7(9), 10952–10962. https://doi.org/10.1021/acsanm.4c01659

Peng, Q. (2023). Materials and prospects of novel solar cells. *E3S Web of Conferences,* 424, 04016. https://doi.org/10.1051/e3sconf/202342404016

Peng, Q., Wang, K., Gong, Y., Zhang, X., Xu, Y., Ma, Y., et al. (2023). Tailoring lignin-derived porous carbon toward high-energy lithium-ion capacitor through varying sp^2- and sp^3-hybridized bonding. *Advanced Functional Materials,* 33(52). https://doi.org/10.1002/adfm.202308284

Prabhurajeshwar, C., Chavan, A., Mallikarjunaiah, N., & Revanasiddappa, K. (2023). Nanomaterials: potential broad-spectrum antimicrobial agents. *Natural Product Communications,* 18(8). https://doi.org/10.1177/1934578x221106904

Prajsner, M. & Winiarz, P. (2024). Application of the electrospinning technique in the preparation of selected electrode materials for solid oxide fuel cells. *Journal of Physics Conference Series,* 2812(1), 012008. https://doi.org/10.1088/1742-6596/2812/1/012008

Prasad, J., Li, S., Zhang, X., Navarro, M., Wang, J., He, W., et al. (2024). Thermodynamic and economic analysis of a trigenerative adiabatic compressed air energy storage. *Journal of Physics Conference Series,* 2929(1), 012003. https://doi.org/10.1088/1742-6596/2929/1/012003

Prasad, M., Hein, S., Danner, T., Prifling, B., Scurtu, R., Hoffmann, A., et al. (2024). Influence of conductive additives and binder on the impedance of lithium-ion battery electrodes: effect of an inhomogeneous distribution. *Journal of the Electrochemical Society,* 171(10), 100518. https://doi.org/10.1149/1945-7111/ad81b9

Praveena, K., Jain, A., Arun, V., Kaliyaperumal, G., Khan, I., Pahwa, S., et al. (2024). A review on next-generation solar solutions: pioneering materials and designs for sustainable energy harvesting. *E3S Web of Conferences,* 505, 02004. https://doi.org/10.1051/e3sconf/202450502004

Preez, H. & Halma, M. (2024). Graphene-based nanomaterials: uses, environmental fate, and human health hazards. *Nano Biomedicine and Engineering,* 16(2), 219–231. https://doi.org/10.26599/nbe.2024.9290059

Pu, C. (2024). Highly ordered silicon/carbon composite materials based on biomass and their application in lithium batteries. *Non-Metallic Material Science,* 6(1), 1–6. https://doi.org/10.30564/nmms.v6i1.6387

Pyun, D., Lee, S., Kim, Y., Jang, G., Choi, D., Jeong, S., et al. (2023). First demonstration of top contact-free perovskite/silicon two-terminal tandem solar cells for overcoming the current density hurdle. *ACS Applied Energy Materials,* 6(5), 2687–2697. https://doi.org/10.1021/acsaem.2c02649

Qiao, R., Shen, X., Mao, C., Di, Y., Zhou, S., Jiang, T., et al. (2024). High-strength silicon anodes with high tap density via compression carbonization for liquid and all-solid-state lithium-ion batteries. *Nano Letters,* 24(49), 15629–15637. https://doi.org/10.1021/acs.nanolett.4c04106

Qiao, Y., Luo, M., Yin, Q., Wang, Y., & Zhou, G. (2023). Molecular dynamics simulation study on thermal transport in graphyne/polyaniline composite system. *Macromolecular Theory and Simulations,* 32(6). https://doi.org/10.1002/mats.202300023

Qiu, Y., Ren, Y., Jia, X., Li, H., & Zhang, M. (2023). Microfluidic construction of polypyrrole-coated core–sheath polyaniline/graphene hybrid fibers with excellent properties for wearable supercapacitors. *ACS Applied Energy Materials,* 6(21), 11189–11198. https://doi.org/10.1021/acsaem.3c02056

Quintero, Y., Herrera, S., Zapata-Benabithe, Z., Manzano-Ramirez, A., & Cruz, L. (2023). Electrochemical performance of activated graphene-based materials for electrodes of double-layer supercapacitors. *ECS Advances,* 2(4), 041001. https://doi.org/10.1149/2754-2734/ad16ba

Ra, H., Lee, S., Jeong, S., Cho, S., & Lee, J. (2023). Advances in heterostructures for optoelectronic devices: materials, properties, conduction mechanisms, device applications. *Small Methods,* 8(2). https://doi.org/10.1002/smtd.202300245

Rafique, S., Burton, M., Badieh, N., Mehraban, S., Tarat, A., Zuo, G., et al. (2023). Ultralow thermal conductivity achieved by all carbon nanocomposites for thermoelectric applications. *Advanced Electronic Materials,* 9(7). https://doi.org/10.1002/aelm.202300023

Rahayu, E., Budiyono, B., & Hadiyanto, H. (2023). The effect of microwave irradiation on the synthesis of graphene from battery waste on capacitance properties. *Periodica Polytechnica Chemical Engineering,* 67(3), 345–354. https://doi.org/10.3311/ppch.21377

Rahim, A., Nadh, R., Saeed, A., Majety, S., Akhil, S., Kumar, N., et al. (2024). Enhanced catalytic, antioxidant, and electrochemical properties of green-synthesized graphene-silver nanocomposite utilizing moringa oleifera leaf extract. *ChemistrySelect,* 9(28). https://doi.org/10.1002/slct.202401848

Rahman, M., Nor, N., Singh, N., Sikiru, S., Dennis, J., Shukur, M., et al. (2023a). One-step solvothermal synthesis by ethylene glycol to produce N-rGO for supercapacitor applications. *Nanomaterials,* 13(4), 666. https://doi.org/10.3390/nano13040666

Rahman, M., Yagyu, J., Islam, M., Fukuda, M., Wakamatsu, S., Tagawa, R., et al. (2023b). Three-dimensional sulfonated graphene oxide proton exchange membranes for fuel cells. *ACS Applied Nano Materials,* 6(3), 1707–1713. https://doi.org/10.1021/acsanm.2c04631

Rai, R., Singh, R., & Guin, M. (2023). Chalcogenide perovskite, an emerging photovoltaic material: current status and future perspectives. *ChemistrySelect,* 8(45). https://doi.org/10.1002/slct.202303550

Rajendhiran, R., Atchudan, R., Jayabal, P., Balasankar, A., Oh, T., Deivasigamani, V., et al. (2023). Prickly pear fruit extract: capping agent for the sol–gel synthesis of discrete titanium dioxide nanoparticles and sensitizer for dye-sensitized solar cell. *Coatings,* 13(3), 579. https://doi.org/10.3390/coatings13030579

Rakshe, D., William, P., Jawale, M., Pawar, A., Korde, S., & Deshpande, N. (2023). Synthesis and characterization of graphene based nanomaterials for energy applications. *Journal of Nano- and Electronic Physics,* 15(3), 030201–030205. https://doi.org/10.21272/jnep.15(3).03020

Raman, S., Sankar, A., & Sindhuja, M. (2023). Advances in silicon nanowire applications in energy generation, storage, sensing, and electronics: a review. *Nanotechnology,* 34(18), 182001. https://doi.org/10.1088/1361–6528/acb320

Ramay, S. & Aldosary, M. (2024). Investigation of structural, morphological, optical and electronic properties of Cu-doped PbS thin films: a comparative experimental and theoretical study. *Physica Scripta,* 99(8), 0859a3. https://doi.org/10.1088/1402-4896/ad6519

Ran, Z. (2023). Application of carbon nanotube composite materials in aviation. *Applied and Computational Engineering,* 23(1), 177–182. https://doi.org/10.54254/2755-2721/23/20230649

Rashid, G., Lone, S., & Mufti, M. (2023). Modeling and performance assessment of an isolated wind-diesel system with flywheel energy storage system. *Wind Engineering,* 47(3), 597–606. https://doi.org/10.1177/0309524x221147601

Rathi, V., Brajpuriya, R., Gupta, R., Parmar, K., & Kumar, A. (2024). Graphene-derived composites: a new frontier in thermoelectric energy conversion. *Energy Advances,* 3(2), 389–412. https://doi.org/10.1039/d3ya00526g

Ravisankar, D., Geetha, D., & Ramesh, P. (2024). Zinc cobalt sulfide with carbon (graphene oxide and CNT) based nanocomposite as positive electrode for asymmetric coin cell supercapacitors. *Physica Scripta,* 99(11), 115995. https://doi.org/10.1088/1402-4896/ad87c6

Rayabharam, A. & Aluru, N. (2024). Modulation of proton permeation through defect engineering in hexagonal boron nitride. *The Journal of Physical Chemistry C,* 128(48), 20528–20536. https://doi.org/10.1021/acs.jpcc.4c03966

Reddygunta, K., Callander, A., Šiller, L., Faulds, K., Berlouis, L., & Ivaturi, A. (2024). Scalable slot-die coated flexible supercapacitors from upcycled pet face shields. *RSC Advances,* 14(18), 12781–12795. https://doi.org/10.1039/d2ra06809e

Reem, A., Almansoob, S., Senan, A., Raj, A., Shah, R., Shrewastwa, M., et al. (2024). *Pseudomonas aeruginosa* and related antibiotic resistance genes as indicators for wastewater treatment. *Heliyon,* 10(9), e29798. https://doi.org/10.1016/j.heliyon.2024.e29798

Ren, S. (2023). Research progress of nano-modified materials for positive electrode of lithium-ion battery. *Applied and Computational Engineering,* 26(1), 158–163. https://doi.org/10.54254/2755-2721/26/20230823

Reza, M., Afroze, S., Кутербеков, К., Kabyshev, A., Bekmyrza, K., Haque, M., et al. (2023). Advanced applications of carbonaceous materials in sustainable water treatment, energy storage, and CO_2 capture: a comprehensive review. *Sustainability,* 15(11), 8815. https://doi.org/10.3390/su15118815

Roberto, S., Andrea, P., & Fernández-Luqueño, F. (2023). Metal oxide nanoparticles (TiO_2, ZnO, and Fe_2O_3) change the functional groups, but not the plant tissue content of common bean plants grown in a greenhouse. *Polish Journal of Environmental Studies,* 32(2), 1343–1352. https://doi.org/10.15244/pjoes/156777

Roohi, Z., Mighri, F., & Zhang, Z. (2024). Conductive polymer-based electrodes and supercapacitors: materials, electrolytes, and characterizations. *Materials,* 17(16), 4126. https://doi.org/10.3390/ma17164126

Roschger, M., Wolf, S., Hasso, R., Genorio, B., Gorgieva, S., & Hacker, V. (2023). Influence of the electrode deposition method of graphene-based catalyst inks for ADEFC on performance. *ACS Applied Materials & Interfaces,* 15(34), 40687–40699. https://doi.org/10.1021/acsami.3c09192

Saber, A., Chen, Y., Mabuti, L., Chaganti, S., Sharma, S., Lüder, J., et al. (2025). Engineering carbonyl-rich conjugated microporous polymers with a pyrene-4,5,9,10-tetraone building block as highly efficient and stable electrodes for energy storage. *Materials Advances*, 6(2), 607–616. https://doi.org/10.1039/d4ma00928b

Sadiq, I., Ali, S., & Ahmad, T. (2023). Graphene-based derivatives heterostructured catalytic systems for sustainable hydrogen energy via overall water splitting. *Catalysts*, 13(1), 109. https://doi.org/10.3390/catal13010109

Saeed, M., Shahzad, U., Rabbee, M., Manzar, R., Al-Humaidi, J., Siddique, A., et al. (2024). Potential development of porous carbon composites generated from the biomass for energy storage applications. *Chemistry - An Asian Journal*, 19(17). https://doi.org/10.1002/asia.202400394

Saleh, H. & Hassan, A. (2023). Synthesis and characterization of nanomaterials for application in cost-effective electrochemical devices. *Sustainability*, 15(14), 10891. https://doi.org/10.3390/su151410891

Sama, A. (2023). Introducing graphene-enhanced technologies to the regulated market to mitigate carbon emissions. *Environmental Progress & Sustainable Energy*, 43(1). https://doi.org/10.1002/ep.14338

Samantaray, S., Mohanty, D., Satpathy, S., & Hung, I. (2024). Exploring recent developments in graphene-based cathode materials for fuel cell applications: a comprehensive overview. *Molecules*, 29(12), 2937. https://doi.org/10.3390/molecules29122937

Sankar, S., Fernandes, J., Santos, M., Espiña, B., Alpuim, P., Díez, A., et al. (2023). Stable graphene membranes for selective ion transport and emerging contaminants removal in water. *Advanced Functional Materials*, 33(23). https://doi.org/10.1002/adfm.202214889

Savitha, H., Kottam, N., Chinnam, S., Madhu, G., & Aishwarya, C. (2024). Recent advances in cost-effective ZnO-based electrode material for lithium-ion batteries. *ChemistrySelect*, 9(37). https://doi.org/10.1002/slct.202402489

Schulte, P., Leso, V., & Iavicoli, I. (2024). Responsible development of emerging technologies. *Journal of Occupational and Environmental Medicine*, 66(7), 528–535. https://doi.org/10.1097/jom.0000000000003100

Seetawan, C., Namhongsa, W., Ruamruk, S., Thaowankaew, S., Zhu, M., Zhang, C., et al. (2024). Graphene aerogel-enhanced DBD reactor with thermoelectric energy recovery for superior plasma activated water production. *Journal of Physics Conference Series*, 2908(1), 012023. https://doi.org/10.1088/1742-6596/2908/1/012023

Sengupta, J. & Hussain, C. (2024). Sensitive and selective detection of heavy metal ions and organic pollutants with graphene-integrated sensing platforms. *Nanoscale*, 16(30), 14195–14212. https://doi.org/10.1039/d4nr00956h

Sezali, N., Ong, H., Villagracia, A., & Hoang, T. (2024). Bio-based nanomaterials for energy application: a review. *Vietnam Journal of Chemistry*, 62(1), 1–12. https://doi.org/10.1002/vjch.202300158

Shadidi, Z., Hamood, R., & Jamal, S. (2024). Electronic properties of armchair and zigzag single-walled carbon nanotubes. *Journal of Science and Technology*, 29(1), 37–44. https://doi.org/10.20428/jst.v29i1.2256

Shah, V., Sinha, R., Cesarski, W., Gao, X., Yuk, S., & Joo, Y. (2024). Modality-tunable exfoliated N-doped graphene as effective electrolyte additive for high-performance lithium–sulfur batteries. *ACS Applied Materials & Interfaces*, 16(40), 53950–53962. https://doi.org/10.1021/acsami.4c12157

Shahazi, R., Majumdar, S., Saddam, A., Mondal, J., Rahman, M., & Alam, M. (2023). Carbon nanomaterials for biomedical applications: a comprehensive review. *Nano Carbons*, 1(1), 448. https://doi.org/10.59400/n-c.v1i1.448

Shahzad, R., Rasul, S., Mamlouk, M., Brewis, I., Shakoor, A., Lukose, C., et al. (2024). Designing molybdenum trioxide and hard carbon architecture for stable lithium-ion battery anodes. *Advanced Materials Interfaces*, 11(31). https://doi.org/10.1002/admi.202400258

Shaker, L., Al-Amiery, A., & Al-Azzawi, W. (2024). Nanomaterials: paving the way for the hydrogen energy frontier. *Discover Nano*, 19(1). https://doi.org/10.1186/s11671-023-03949-8

Shang, M., Bu, S., Hu, Z., Zhao, Y., Liao, J., Zheng, C., et al. (2024). Polyacrylonitrile as an efficient transfer medium for wafer-scale transfer of graphene. *Advanced Materials*, 36(29). https://doi.org/10.1002/adma.202402000

Shanmuga, P.S. & Suseem, S.R. (2024). Plant-based carbon dots are a sustainable alternative to conventional nanomaterials for biomedical and sensing applications. *Nano Express*, 5(1), 012002. https://doi.org/10.1088/2632-959x/ad100c

Shen, N., Lai, Q., Chen, N., Pang, Y., Chen, H., Zhang, W., et al. (2023). Hard carbon anode synthesized by an in situ porous strategy for advanced potassium-ion batteries. *ACS Sustainable Chemistry & Engineering*, 11(39), 14572–14581. https://doi.org/10.1021/acssuschemeng.3c04086

Shen, Y., Wen, J., Wang, T., Yang, G., Ma, H., Lin, M., et al. (2024). Surface N sites mediate the formation of Li metal cluster@N-enriched hierarchically porous carbon for ultrahigh-capacity lithium storage. *ACS Applied Materials & Interfaces*, 16(18). https://doi.org/10.1021/acsami.4c00416

Sheng, O., Jin, C., Yang, T., Ju, Z., Luo, J., & Tao, X. (2023). Designing biomass-integrated solid polymer electrolytes for safe and energy-dense lithium metal batteries. *Energy & Environmental Science*, 16(7), 2804–2824. https://doi.org/10.1039/d3ee01173a

Shi, M., Ren, Y., Cao, J., Kuang, Z., Zhuo, X., & Xie, H. (2023). Current situation and development prospects of discharge pretreatment during recycling of lithium-ion batteries: a review. *Batteries & Supercaps*, 7(2). https://doi.org/10.1002/batt.202300477

Shi, X., Liu, J., Zhang, H., Xue, Z., Zhao, Z., Zhang, Y., et al. (2024). Solid electrolyte interphase recombination on graphene nanoribbons for lithium anode. *ACS Nano*, 18(12), 8827–8838. https://doi.org/10.1021/acsnano.3c11796

Shitanda, I., Sugaya, K., Baba, C., Loew, N., Yamagata, Y., Miyamoto, K., et al. (2023). Rheo-impedance measurements for the dispersibility evaluation of electrode slurries. *ACS Applied Electronic Materials*, 5(8), 4394–4400. https://doi.org/10.1021/acsaelm.3c00612

Shukla, N., Verma, A., & Tiwari, S. (2023). Optimization of efficient perovskite-Si hybrid tandem solar cells. *Material Science Research India*, 20(1), 25–40. https://doi.org/10.13005/msri/200104

Si, H., Ma, J., Xiao, X., Wang, Q., Geng, S., & Fu, L. (2025). Solid-state sodium-ion batteries: theories, challenges and perspectives. *Chemistry - A European Journal*, 31(5). https://doi.org/10.1002/chem.202403247

Siddiqa, A., Yhobu, Z., Nagaraju, D., Padaki, M., Budagumpi, S., Pasupuleti, V., et al. (2023). Review and perspectives of sustainable lignin, cellulose, and lignocellulosic carbon special structures for energy storage. *Energy & Fuels*, 37(4), 2498–2519. https://doi.org/10.1021/acs.energyfuels.2c03557

Silva, R., Abeykoon, A., & Kottegoda, I. (2023). A review on synthesis of graphene-based materials for energy storage devices. *Sri Lankan Journal of Physics*, 24(2), 135–159. https://doi.org/10.4038/sljp.v24i2.8132

Simpa, P., Solomon, N., Adenekan, O., & Obasi, S. (2024). Nanotechnology's potential in advancing renewable energy solutions. *Engineering Science & Technology Journal*, 5(5), 1695–1710. https://doi.org/10.51594/estj.v5i5.1137

Sinha, N. & Sinha, P. (2024). Recent advancements in nano-enhanced solar technologies. *Journal of Mountain Research*, 19(1). https://doi.org/10.51220/jmr.v19-i1.59

Singh, H., Wu, J., Lagemann, K., & Nath, M. (2024a). Highly efficient dopamine sensing with a carbon nanotube-encapsulated metal chalcogenide nanostructure. *ACS Applied Nano Materials*, 7(5), 4814–4823. https://doi.org/10.1021/acsanm.3c05422

Singh, K., Trabelsi, Y., Mahajan, P., & Khanna, V. (2024b). Role of graphene in next-generation electrochemical energy storage systems: insights and innovations. *Journal of the Electrochemical Society*, 171(10), 100501. https://doi.org/10.1149/1945-7111/ad7bee

Singh, M., Akash, A., & Tiwari, J. (2024c). Solar cells based on Pb-free and perovskite-inspired materials: futuristic of perovskite solar cells. *ACS Applied Energy Materials*, 7(22), 10212–10229. https://doi.org/10.1021/acsaem.4c02047

Song, G., Lee, S., Kim, T., Jung, M., Kim, K., Choi, S., et al. (2024). Mechano-electrochemical behavior of nanostructured Li- and Mn-rich layered oxides with superior capacity retention and voltage decay for sulfide-based all-solid-state batteries. *Advanced Energy Materials*, 14(47). https://doi.org/10.1002/aenm.202403374

Song, G., Shi, Y., Jiang, S., & Pang, H. (2023). Recent progress in MOF-derived porous materials as electrodes for high-performance lithium-ion batteries. *Advanced Functional Materials*, 33(42). https://doi.org/10.1002/adfm.202303121

Song, J., Xu, X., & Liang, X. (2024). Thermal transport properties of graphene aerogel as an advanced carrier for enhanced energy storage. *Physical Chemistry Chemical Physics*, 26(3), 2025–2034. https://doi.org/10.1039/d3cp05078e

Song, X., Huang, W., Ge, R., Wang, H., & Xing, J. (2024). Electrospun nanofibers constituted of a polyimide core decorated with polyhedral oligomeric silsesquioxanes and a TiO_2 shell as separators for lithium-ion batteries. *ACS Applied Nano Materials*, 7(18), 21565–21577. https://doi.org/10.1021/acsanm.4c03481

Song, Y., Bae, W., Ahn, J., Son, Y., Kwon, M., Kwon, C., et al. (2023). Carbon nanocluster-mediated nanoblending assembly for binder-free energy storage electrodes with high capacities and enhanced charge transfer kinetics. *Advanced Science*, 10(22). https://doi.org/10.1002/advs.202301248

Srivishnu, K., Markapudi, P., Sundaram, S., & Giribabu, L. (2023). Semitransparent perovskite solar cells for building integrated photovoltaics: recent advances. *Energies*, 16(2), 889. https://doi.org/10.3390/en16020889

Stetsenko, M. & Abaszade, R. (2023). X-ray phase analysis of carbon nanotubes obtained by the arc discharge method. *UNEC J. Eng. Appl. Sci.*, 3(1), 15–20. https://doi.org/10.61640/ujeas.2023.0503

Su, D., Zhang, H., Zhang, J., & Zhao, Y. (2023). Design and synthesis strategy of MXenes-based anode materials for sodium-ion batteries and progress of first-principles research. *Molecules*, 28(17), 6292. https://doi.org/10.3390/molecules28176292

Su, Z., Huang, J., & Wu, L. (2024). Economic study of wind and solar power generation with energy storage configuration. *Highlights in Science Engineering and Technology*, 112, 445–453. https://doi.org/10.54097/jh34fy18

Suci, W., Jamaluddin, A., Panuntun, A., Stulasti, K., Setyawati, R., & Azinuddin, Y. (2023). The effect of activated carb on derived from black betel leaf biomass waste as composite anodes on lithium-ion battery applications. *E3S Web of Conferences*, 465, 01022. https://doi.org/10.1051/e3sconf/202346501022

Sui, H. (2023). Research on preparation and progress of two-dimensional perovskite solar cells. *Academic Journal of Science and Technology*, 5(2), 196–199. https://doi.org/10.54097/ajst.v5i2.6865

Sulaiman, S. & Wealth, L. (2025). Synthesis and characterization of reduced graphene oxide for thermoelectric materials using hummers' method. *Journal of Physics Conference Series*, 2933(1), 012010. https://doi.org/10.1088/1742-6596/2933/1/012010

Sun, C., Han, Z., Wang, X., Liu, B., Li, Q., Li, H., et al. (2023). Advanced carbons nanofibers-based electrodes for flexible energy storage devices. *Advanced Functional Materials*, 33(52). https://doi.org/10.1002/adfm.202305606

Sun, H., Li, J., Liang, W., Gong, X., Jing, A., Yang, W., et al. (2023). Porous organic polymers as active electrode materials for energy storage applications. *Small Methods*, 8(8). https://doi.org/10.1002/smtd.202301335

Sun, K., Ying, S., Fang, T., Zhou, G., & Liu, X. (2024). Revealing the two-stage charging process in sulfuric acid electrolyte by molecular dynamics simulation. *Langmuir*, 40(29), 15080–15091. https://doi.org/10.1021/acs.langmuir.4c01431

Sun, X. (2023). Application and performance evaluation of solid state batteries in renewable energy storage systems. *TETR*, 1, 22–26. https://doi.org/10.62051/7r9hk854

Sun, Y., He, S., Li, Z., Mclaughlin, A., Chen, K., Shao, Z., et al. (2024). Development of nanostructured lanthanum strontium cobalt ferrite/gadolinian-doped ceria composite electrodes of solid oxide cells formed by in situ polarization. *ACS Applied Materials & Interfaces*, 16(17), 21818–21827. https://doi.org/10.1021/acsami.4c01223

Sun, Z., Zhao, C., Qian, Y., Xing, J., & Niu, J. (2024). Cavern spacing of CAES cavern group in hard rock based on plastic zone and deformation analysis. *IOP Conference Series Earth and Environmental Science*, 1335(1), 012044. https://doi.org/10.1088/1755-1315/1335/1/012044

Sundaram, G., El-Mahdy, A., Pham, P., Kunjiappan, S., & Kumar, A. (2024). Synergistic innovations: organometallic frameworks on graphene oxide for sustainable eco-energy solutions. *Chemengineering*, 8(3), 61. https://doi.org/10.3390/chemengineering8030061

Sundhar, A., Bejigo, K., & Kim, S. (2023). Scope and significance of transition metal oxide nanomaterials for next-generation li-ion batteries. *Materials Chemistry Frontiers*, 7(20), 4613–4634. https://doi.org/10.1039/d3qm00226h

Syed, Q. & Choudhury, J. (2023). Statistical analysis and employability skills equation for graduate students in the Bahamas. *Annals of Social Sciences & Management Studies*, 9(3). https://doi.org/10.19080/asm.2023.09.555765

Talukder, N., Wang, Y., Nunna, B., & Lee, E. (2024). N-doped graphene (N-G)/MOF(ZIF-8)-based/derived materials for electrochemical energy applications: synthesis, characteristics, and functionality. *Batteries*, 10(2), 47. https://doi.org/10.3390/batteries10020047

Tamuly, J., Bhattacharjya, D., & Saikia, B. (2023). Low-grade coal feedstock for the synthesis of high-valued graphene derivatives for energy storage application. *Energy & Fuels*, 37(4), 3260–3271. https://doi.org/10.1021/acs.energyfuels.2c04086

Tan, S., Yang, H., Zhang, Z., Xu, X., Xu, Y., Zhou, J., et al. (2023). The progress of hard carbon as an anode material in sodium-ion batteries. *Molecules*, 28(7), 3134. https://doi.org/10.3390/molecules28073134

Tănase, N., Ilie, C., Popa, M., Demeter, L., Ovezea, D., & Daniel, L. (2023). Inertial energy storage system, applicative extension. International *Journal of Mechatronics and Applied Mechanics*, I(13). https://doi.org/10.17683/ijomam/issue13.12

Tang, Y., He, J., Peng, J., Yang, J., Wu, Z., Liu, P., et al. (2024). Electrochemical behavior of the biomass hard carbon derived from waste corncob as a sodium-ion battery anode. *Energy & Fuels*, 38(8), 7389–7398. https://doi.org/10.1021/acs.energyfuels.4c00564

Tao, Z., Zhao, Y., Wang, Y., & Zhang, G. (2024). Recent advances in carbon nanotube technology: bridging the gap from fundamental science to wide applications. *C – Journal of Carbon Research*, 10(3), 69. https://doi.org/10.3390/c10030069

Tariq, H., Awan, S., Hussain, D., Rizwan, S., Shah, S., Zainab, S., et al. (2023). Enhancing supercapacitor performance through design optimization of laser-induced graphene and MWCNT coatings for flexible and portable energy storage. *Scientific Reports*, 13(1). https://doi.org/10.1038/s41598-023-48518-2

Tene, T., Bellucci, S., Guevara, M., Romero, P., Guapi, A., Gahramanli, L., et al. (2024). Role of graphene oxide and reduced graphene oxide in electric double-layer capacitors: a systematic review. *Batteries*, 10(7), 256. https://doi.org/10.3390/batteries10070256

Teng, Y., Li, J., Yao, J., Kang, L., Li, Q., Ren, Y., et al. (2023). Filled carbon-nanotube heterostructures: from synthesis to application. *Microstructures*. https://doi.org/10.20517/microstructures.2023.07

Thate, A., Pakhare, K., Patil, S., & Bhuse, V. (2023). Environmentally benign synthesis of TiO_2-ZnO nanocomposite for efficient dye-sensitized solar cell. *Materials Research Express*, 10(5), 055001. https://doi.org/10.1088/2053-1591/acd23c

Thirugnanasambandan, T., Karuppaiah, C., Sankar, B., & Gopinath, S. (2024). Insights and potentials of two-dimensional black phosphorous-based solar cells. *Physica Scripta*, 99(5), 052002. https://doi.org/10.1088/1402-4896/ad3a27

Tian, C. (2024). Application of metal-based nanomaterials in lithium batteries. *Applied and Computational Engineering*, 59(1), 22–29. https://doi.org/10.54254/2755-2721/59/20240742

Topalović, Z. & Haas, R. (2024). Role of renewables in energy storage economic viability in the western Balkans. *Energies*, 17(4), 955. https://doi.org/10.3390/en17040955

Tron, A., Paolella, A., & Beutl, A. (2023). New insights of infiltration process of argyrodite Li_6PS_5Cl solid electrolyte into conventional lithium-ion electrodes for solid-state batteries. *Batteries*, 9(10), 503. https://doi.org/10.3390/batteries9100503

Tsai, L. (2024). Research progress of carbon nanomaterials for lithium-ion battery anodes. *Applied and Computational Engineering*, 59(1), 155–160. https://doi.org/10.54254/2755-2721/59/20240788

Tyagi, S. & Negi, S. (2023). Calculation of density of states of pristine and functionalized carbon nanotubes: a DFT approach. *Indian Journal of Science and Technology*, 16(40), 3567–3574. https://doi.org/10.17485/ijst/v16i40.1019

Ubhi, M., Kaur, M., Grewal, J., & Sharma, V. (2023). Phosphorous- and boron-doped graphene-based nanomaterials for energy-related applications. *Materials*, 16(3), 1155. https://doi.org/10.3390/ma16031155

Ugwuanyi, E., Nwokediegwu, Z., Dada, M., Majemite, M., & Obaigbena, A. (2024). The impact of nanomaterials in enhancing wastewater treatment processes: a review. *Magna Scientia Advanced Research and Reviews*, 10(1), 273–285. https://doi.org/10.30574/msarr.2024.10.1.0030

Untarabut, P. & Banlusan, K. (2023). Atomistic insight into the role of porous structure in accommodating the volume expansion of silicon anodes for lithium-ion batteries. *Journal of the Electrochemical Society*, 170(11), 110512. https://doi.org/10.1149/1945-7111/ad0757

Varatharajan, P., Banu, I., Mamat, M., & Vasimalai, N. (2023). Electrochemical analysis of asymmetric supercapacitors based on $BiCoO_3$@g-C_3N_4 nanocomposites. *Dalton Transactions*, 52(38), 13704–13715. https://doi.org/10.1039/d3dt01758c

Vengadesan, K., Madaswamy, S., Natarajan, V., & Dhanusuraman, R. (2023). Biocarbon derived from seeds of palmyra palm tree for a supercapacitor application. *Advanced Nano Research*, 6(1), 1–10. https://doi.org/10.21467/anr.6.1.1-10

Visser, E., Seroka, N., & Khotseng, L. (2023). Catalytic properties of biochar as support material potential for direct methanol fuel cell: a review. *ACS Omega*, 8(44), 40972–40981. https://doi.org/10.1021/acsomega.3c02283

Vozniakovskii, A., Smirnova, E., Апраксин, Р., Кидалов, С., & Voznyakovskiĭ, A. (2023). Use of few-layer graphene synthesized under conditions of self-propagating high-temperature synthesis for supercapacitors applications. *Nanomaterials*, 13(16), 2368. https://doi.org/10.3390/nano13162368

Wang, C., Li, W., Jin, Y., Liu, J., Wang, H., & Zhang, Q. (2023). Functional separator enabled by covalent organic frameworks for high-performance li metal batteries. *Small*, 19(28). https://doi.org/10.1002/smll.202300023

Waris, M., Khan, R., Afzal, A., Iqbal, M., Imran, M., Mumtaz, M., et al. (2023). Composite electrode materials based on nickel cobalt sulfide/carbon nanotubes to enhance the redox activity for high performance asymmetric supercapacitor devices. *Physica Scripta*, 98(10), 105979. https://doi.org/10.1088/1402-4896/acf52f

Wei, B., Chen, C., Xiao, F., & Hong, F. (2024a). Distributed coordinated speed control of flywheel energy storage matrix systems with model uncertainties and disturbances. International *Journal of Robust and Nonlinear Control*, 34(14), 9797–9813. https://doi.org/10.1002/rnc.7495

Wei, C., Liu, C., Xiao, Y., Wu, Z., Luo, Q., Jiang, Z., et al. (2024b). SnF_2-induced ultifunctional interface-stabilized $Li_{5.5}PS_{4.5}Cl_{1.5}$-based all-solid-state lithium metal batteries. *Advanced Functional Materials*, 34(18). https://doi.org/10.1002/adfm.202314306

Wei, L., Liu, T., Zhang, Y., Liu, H., & Ge, L. (2024c). Taurine-functionalized carbon nanotubes as electrode catalysts for improvement in the performance of vanadium redox flow battery. *Catalysts*, 14(4), 281. https://doi.org/10.3390/catal14040281

Wei, L., Zhou, Z., Wang, B., & Fang, F. (2023). ADRC-based control strategy for DC-link voltage of flywheel energy storage system. *Energy Science & Engineering*, 11(11), 4141–4154. https://doi.org/10.1002/ese3.1569

Wei, P., Li, P., Zia, M., Bhuiyan, S., Liu, Y., Chen, D., et al. (2024d). A comparative review on carbon nanomaterials and their impact on plant growth with the lens of green chemistry principles. *Global Challenges*, 8(11). https://doi.org/10.1002/gch2.202400008

Wei, X., Zhao, T., Yang, Y., Shi, M., Wang, J., Jin, Z., et al. (2024e). Design of moisture-enabled electric generators utilizing sp- and sp^2-hybridized two-dimensional carbon materials: a minireview and perspectives. *Energy & Fuels*, 38(21), 20242–20257. https://doi.org/10.1021/acs.energyfuels.4c03934

Wen, J., Zhao, Y., Wu, P., Liu, Y., Zheng, X., Lin, R., et al. (2023). Heterojunction formed via 3D-to-2D perovskite conversion for photostable wide-bandgap perovskite solar cells. *Nature Communications*, 14(1). https://doi.org/10.1038/s41467-023-43016-5

Weng, S., Yang, G., Zhang, S., Liu, X., Zhang, X., Liu, Z., et al. (2023). Kinetic limits of graphite anode for fast-charging lithium-ion batteries. *Nano-Micro Letters*, 15(1). https://doi.org/10.1007/s40820-023-01183-6

Williams, A., Moore, E., Thomas, A., & Johnson, J. (2023). Graphene-based materials in dental applications: antibacterial, biocompatible, and bone regenerative properties. *International Journal of Biomaterials*, 2023, 1–18. https://doi.org/10.1155/2023/8803283

Wittmann, S., Pindl, S., Sawallich, S., Nagel, M., Michalski, A., Pandey, H., et al. (2023). Assessment of wafer-level transfer techniques of graphene with respect to semiconductor industry requirements. *Advanced Materials Technologies*, 8(8). https://doi.org/10.1002/admt.202201587

Wojciechowska, O., Costabile, A., & Kujawska, M. (2023). The gut microbiome meets nanomaterials: exposure and interplay with graphene nanoparticles. *Nanoscale Advances*, 5(23), 6349–6364. https://doi.org/10.1039/d3na00696d

Wu, F., Dong, Y., Su, Y., Wei, C., Chen, T., Yan, W., et al. (2023). Benchmarking the effect of particle size on silicon anode materials for lithium-ion batteries. *Small*, 19(42). https://doi.org/10.1002/smll.202301301

Wu, X. (2024). Analysis of cancer detection methods based on nanomaterials. *Applied and Computational Engineering*, 85(1), 95–100. https://doi.org/10.54254/2755-2721/85/20240799

Wu, Y., Zhang, H., Xu, Y., Tang, Z., & Li, Z. (2024). Ferroelectric $BiFeO_3$ modified PVDF-based electrolytes for high-performance lithium metal batteries. *Journal of Materials Chemistry A*, 12(31), 20403–20413. https://doi.org/10.1039/d4ta03225j

Wu, Z. & Liang, X. (2023). Solar cell performance change based on polymer change. *Journal of Physics Conference Series*, 2598(1), 012015. https://doi.org/10.1088/1742-6596/2598/1/012015

Xia, S., Song, J., Zhou, Q., Liu, L., Ye, J., Wang, T., et al. (2023). A separator with double coatings of $Li_4Ti_5O_{12}$ and conductive carbon for Li-S battery of good electrochemical performance. *Advanced Science*, 10(22). https://doi.org/10.1002/advs.202301386

Xiao, W., Wu, X., Yang, S., Zha, Y., & Liu, S. (2024). Design of composite N-doped carbon nanofiber/TiO_2/diatomite separator for lithium–sulfur batteries. *Materials*, 17(22), 5615. https://doi.org/10.3390/ma17225615

Xiao, Y., Pang, Y., Yan, Y., Qian, P., Zhao, H., Manickam, S., et al. (2023). Synthesis and functionalization of graphene materials for biomedical applications: recent advances, challenges, and perspectives. *Advanced Science*, 10(9). https://doi.org/10.1002/advs.202205292

Xie, S., Lei, M., Sun, J., Yang, C., Liu, W., Yun, D., et al. (2024). Corrosion behavior of 20G and TP347H in Molten LiCl-NaCl-KCl salt. *Nanomaterials*, 14(12), 1026. https://doi.org/10.3390/nano14121026

Xiong, S., Xu, X., Jiao, X., Wang, Y., Kapitanova, O., Song, Z., et al. (2023). Mechanical failure of solid-state electrolyte rooted in synergy of interfacial and internal defects. *Advanced Energy Materials*, 13(14). https://doi.org/10.1002/aenm.202203614

Xu, C., Yuan, L., Zhang, R., Cheng, J., & Yu, H. (2024). Ethylene carbonate-free electrolytes for high voltage lithium-ion batteries: progress and perspectives. *Energy & Fuels*, 38(19), 18208–18226. https://doi.org/10.1021/acs.energyfuels.4c02728

Xu, J., Li, C., Yu, S., Wang, Z., Pritzel, C., Jiang, X., et al. (2023). Flexible carbon cloth supercapacitors with high power and energy densities. *Energy Technology*, 11(8). https://doi.org/10.1002/ente.202300214

Yadav, A., Yadav, K., Ahmad, R., & Abd–Elsalam, K. (2023). Emerging frontiers in nanotechnology for precision agriculture: advancements, hurdles and prospects. *Agrochemicals*, 2(2), 220–256. https://doi.org/10.3390/agrochemicals2020016

Yadav, S. & Daniel, S. (2023). Green synthesis of zero-dimensional carbon nanostructures in energy storage applications—a review. *Energy Storage*, 6(1). https://doi.org/10.1002/est2.500

Yang, D. & Liao, N. (2024). Interfacial compatibility design of high-performance electrolytes for SiCO-based all-solid-state lithium battery. *Journal of Physics Conference Series*, 2783(1), 012044. https://doi.org/10.1088/1742-6596/2783/1/012044

Yanılmaz, M., Abdolrazzaghian, E., Chen, L., Atıcı, B., & Kim, J. (2023). Flexible centrifugally spun PVP based SnO_2@carbon nanofiber electrodes. *Apl Materials*, 11(9). https://doi.org/10.1063/5.0163981

Yao, Z., Zhang, J., Yang, D., Zhang, D., Yang, B., & Liang, F. (2023). Achieving dendrite-free solid-state lithium–metal batteries via in situ construction of Li_3P/LiCl interfacial layers. *ACS Applied Materials & Interfaces*, 16(1), 869–877. https://doi.org/10.1021/acsami.3c16118

Yarlagadda, V. & Ramaswamy, N. (2025). How good is graphene carbon as a fuel cell catalyst support? *Journal of the Electrochemical Society*, 172(1), 014503. https://doi.org/10.1149/1945-7111/ada642

Yaşa, S., Kumbasi, O., Arvas, M., Gençten, M., Şahin, M., & Şahin, Y. (2023). S-doped graphene oxide/N-doped graphene oxide/PANI: a triple composite for high-performance supercapacitor applications. *ECS Journal of Solid State Science and Technology*, 12(5), 051002. https://doi.org/10.1149/2162-8777/acd3af

Yasmin, R., Amin, B., Shah, R., & Barton, A. (2024). A survey of commercial and industrial demand response flexibility with energy storage systems and renewable energy. *Sustainability*, 16(2), 731. https://doi.org/10.3390/su16020731

Ye, J., Lu, J., & Wen, D. (2023). Engineering carbon nanomaterials toward high-efficiency bioelectrocatalysis for enzymatic biofuel cells: a review. *Materials Chemistry Frontiers*, 7(22), 5806–5825. https://doi.org/10.1039/d3qm00615h

Ye, L., Qi, S., Cheng, T., Jiang, Y., Feng, Z., Wang, M., et al. (2024). Vanadium redox flow battery: review and perspective of 3D electrodes. *ACS Nano*, 18(29), 18852–18869. https://doi.org/10.1021/acsnano.4c06675

Yerdauletov, M., Nazarov, K., Muxametuli, B., Yeleuov, M., Daulbayev, C., Abdulkarimova, R., et al. (2023). Characterization of activated carbon from rice husk for enhanced energy storage devices. *Molecules*, 28(15), 5818. https://doi.org/10.3390/molecules28155818

Yi, M., Cui, Z., & Manthiram, A. (2024). Impact of electrolyte on direct-contact prelithiation of silicon-graphite anodes in lithium-ion cells with high-nickel cathodes. *ACS Applied Materials & Interfaces*, 16(32), 42270–42282. https://doi.org/10.1021/acsami.4c08929

Yi, Y., Chang, L., Wu, B., Zhao, J., Hu, P., Li, L., et al (2024). Life cycle assessment of energy storage technologies for new power systems under dual-carbon target: a review. *Energy Technology*, 12(5). https://doi.org/10.1002/ente.202301129

Yılmazoğlu, E. & Karakuş, S. (2023). Recent developments in nanostructured materials for supercapacitor electrodes. *Journal of the Turkish Chemical Society Section a Chemistry*, 10(4), 1107–1122. https://doi.org/10.18596/jotcsa.1320655

Yin, L., Xie, D., Gong, K., Shi, C., Qian, X., & Zhou, K. (2023a). Phase change materials encapsulated in coral-inspired organic–inorganic aerogels for flame-retardant and thermal energy storage. *ACS Applied Nano Materials*, 6(10), 8752–8762. https://doi.org/10.1021/acsanm.3c01128

Yin, Y., Liu, Q., Zhao, Y., Chen, T., Wang, J., Gui, L. et al. (2023b). Recent progress and future directions of biomass-derived hierarchical porous carbon: designing, preparation, and supercapacitor applications. *Energy & Fuels*, 37(5), 3523–3554. https://doi.org/10.1021/acs.energyfuels.2c04093

Yoon, T., Park, W., You, J., & Na, S. (2024). Investigation of direct electron transfer of glucose oxidase on a graphene-CNT composite surface: a molecular dynamics study based on electrochemical experiments. *Nanomaterials*, 14(13), 1073. https://doi.org/10.3390/nano14131073

Younas, A., Rashad, M., Rafaqat, A., Maqbool, H., Gul, K., Ahmad, M., Shahbaz, J., Ku, K., & Umair, M. (2024). Adsorption of organic compounds and hazardous microorganisms from sewage and surfactant-containing wastewater using carbon-based nanomaterials: a focus on graphene oxide and carbon nanotube. *Biological and Clinical Sciences Research Journal*, 2024(1), 1052. https://doi.org/10.54112/bcsrj.v2024i1.1052

Yousf, N., Ouda, E., & Abdalla, F. (2024). Electrochemical enhancement of carbon nanotubes utilizing MnO_2 for energy storage applications. *Fayoum University Journal of Engineering*, 7(2), 120–126. https://doi.org/10.21608/fuje.2024.343801

Yu, T., He, W., Zhang, Q., & Ma, D. (2023). Advanced nanomaterials and characterization techniques for photovoltaic and photocatalysis applications. *Accounts of Materials Research*, 4(6), 507–521. https://doi.org/10.1021/accountsmr.3c00012

Yu, Y., Zhang, Y., Wang, H., & Chen, G. (2024). Emerging molten salts based electrochemical energy technologies. *Electrochemistry*, 92(4), 043003–043003. https://doi.org/10.5796/electrochemistry.24-69009

Yuan, G., Liu, J., & Zhang, Y. (2024). Sub-1 nm nanomaterials applied in energy storage. *Batteries & Supercaps*, 7(4). https://doi.org/10.1002/batt.202300623

Yuan, S., Wang, J., Dong, B., & Dai, X. (2023). Biomass-derived carbonaceous materials with graphene/graphene-like structures: definition, classification, and environmental applications. *Environmental Science & Technology*, 57(45), 17169–17177. https://doi.org/10.1021/acs.est.3c04203

Zafar, H., Zainab, S., Masood, M., Sohail, M., Shah, S., Karim, M., O'Mullane, A., Ostrikov, K., Will, G., & Wahab, M. (2023). Recent advances on nitrogen-doped porous carbons towards electrochemical supercapacitor applications. *The Chemical Record*, 24(1). https://doi.org/10.1002/tcr.202300161

Zafar, M. & Jacob, M. (2024). Instant upcycling of microplastics into graphene and its environmental application. *Small Science*, 4(10). https://doi.org/10.1002/smsc.202400176

Zafar, S., Cherakkara, A., Misnon, I., Yang, C., & Jose, R. (2024). Graphene-like materials from biomass using deep eutectic solvents: a review. *ACS*

Sustainable Resour. Manage., 1(9), 2047–2073. https://doi.org/10.1021/acssusresmgt.4c00183

Zeng, X., Yue, H., Wu, J., Chen, C., & Liu, L. (2023a). Hybrid ionically covalently cross-linked network binder for high-performance silicon anodes in lithium-ion batteries. *Batteries,* 9(5), 276. https://doi.org/10.3390/batteries9050276

Zeng, Z., Mao, Y., Hu, Z., Chen, K., Huang, Q., Song, Y., Wu, Z., Zhang, P., Chen, T., & Guo, X. (2023b). Research progress and commercialization of biologically derived hard carbon anode materials for sodium-ion batteries. *Industrial & Engineering Chemistry Research,* 62(38), 15343–15359. https://doi.org/10.1021/acs.iecr.3c00818

Zhang, H. (2024). Graphene-based materials in energy storage and conversion devices. *Science and Technology of Engineering, Chemistry and Environmental Protection,* 1(9). https://doi.org/10.61173/fsmn7x52

Zhao, D., Yang, Y., Zhang, C., & Zhu, G. (2023a). Effect of Nb doping on phase, microstructure and lithium-ion conductivity of $Li_7La_3Zr_2O_{12}$ solid electrolyte. *Engineering Advances,* 3(2), 84–88. https://doi.org/10.26855/ea.2023.04.001

Zhao, J., Huang, J., Chen, Y., & Chen, J. (2023b). Slope-dominated N/S co-doped carbon anode derived from pitch for sodium-ion batteries. *ACS Applied Engineering Materials,* 1(10), 2524–2532. https://doi.org/10.1021/acsaenm.3c00349

Zhao, J., Wang, F., Ruan, Q., Wu, Y., Zhang, B., & Lü, Y. (2024). Hybrid energy storage systems for fast-developing renewable energy plants. *Journal of Physics Energy,* 6(4), 042003. https://doi.org/10.1088/2515-7655/ad6fd4

Zhao, L., Chang, Y., Qiu, S., Liu, H., Zhao, J., & Gao, J. (2023c). High mechanical energy storage capacity of ultranarrow carbon nanowires bundles by machine learning driving predictions. *Advanced Energy and Sustainability Research,* 4(11). https://doi.org/10.1002/aesr.202300112

Zhou, S., Li, M., Wang, P., Cheng, L., Chen, L., Huang, Y., Yu, S., Mo, F., & Wei, J. (2023). Printed solid-state batteries. *Electrochemical Energy Reviews,* 6(1). https://doi.org/10.1007/s41918-023-00200-x

Zhou, Y., Wang, H., & Wang, J. (2024). Composite graphene-modified aluminum foil cathode current collectors for lithium-ion battery with enhanced mechanical and electrochemical performances. *Batteries & Supercaps,* 7(6). https://doi.org/10.1002/batt.202400028

Index

For Product Safety Concerns and Information please contact our
EU representative GPSR@taylorandfrancis.com Taylor & Francis
Verlag GmbH, Kaufingerstraße 24, 80331 München, Germany